P9-APJ-336

REACTIONS AND PROPERTIES OF MONOMERS AND POLYMERS

REACTIONS AND PROPERTIES OF MONOMERS AND POLYMERS

ALBERTO D'AMORE
GENNADY ZAIKOV
EDITORS

Nova Science Publishers, Inc.
New York

For permission to use material from this book please contact us:
Telephone 631-231-7269; Fax 631-231-8175
Web Site: http://www.novapublishers.com

LIBRARY OF CONGRESS CATALOGING-IN-PUBLICATION DATA
Monomers and polymers : reactions and properties / Alberto D'Amore and Gennady Zaikov, editor[s].
p. cm.
Includes index.
ISBN 13 978-1-60021-415-8
ISBN 10 1-60021-415-0
1. Monomers--Research. 2. Polymers--Research. 3. Polymerization--Research. 4. Chemical reactions--Research. 5. Chemical processes--Research. I. D'Amore, Alberto. II. Zaikov, Gennadii Efremovich.
QD281.P6M666 2006
547'.7--dc22 2006025951

Published by Nova Science Publishers, Inc. ✦ New York

CONTENTS

PREFACE

"I do not state my new scientific ideas anymore
and I do not write and publish scientific articles
in order to not pollute an environment"
Prof. Victor Shlyapintokh
Russian Academy of Sciences
Moscow, 2005

"In time some people are getting wiser
and some are just getting older"
Prof. Victor Manuel de Matos Lobo
Coimbra University
Portugal, 2004

The given collection of articles has been prepared in view of those wise ideas which have been stated by two world-recognized scientists from Portugal and Russia and are cited as epigraphs of the preface.

This collection of articles contains reviews and original papers of scientists from Italy, Spain, Bulgaria, Georgia and Russia. These articles have been specially ordered by the editor for this book.

This book is mainly dedicated to the processes of ozonolysis, destruction and stabilization of polymers, reactions and structure formation of polymers in fractal spaces, hydrosilylation reactions, antiadhesive coatings, synergism and antagonism in processes of polymers phototransformation and light stabilization, selective oxidation, carbocations reactions and polymer analogous reactions.

Editors of the collection (as well as contributors of this volume) will be glad to receive readers' feedback so that it could be considered in the further work.

Prof. Alberto D'Amore
Second University of Naples (SUN), Aversa (CE), Italy

Prof. Gennady E. Zaikov
Institute of Biochemical Physics
Russian Academy of Sciences, Moscow, Russia

In: Reactions and Properties of Monomers and Polymers
Editors: A. D'Amore and G. Zaikov, pp. 1-20
ISBN: 1-60021-415-0

Chapter 1

HYDROSILYLATION REACTIONS OF METHYLHYDRIDESILOXANE TO ACRYLATES AND METHACRYLATES

O. Mukbaniani*, N. Pirtskheliani, T. Tatrishvili and N. Mukbaniani

I. Javakhishvili Tbilisi State University, I. Chavchavadze Ave.,1, 380028 Tbilisi, Georgia

ABSTRACT

The reaction of hydrosilylation of α,ω–bis(trimethylsiloxy)methylhydridesiloxane (n≈35) with acrylate and methacrylate at various ratio of initial compounds, in the presence of platinum hydrochloric acid, or platinum on the carbon Pt/C has been investigated. It was established that in the presence of platinum catalysts, hydrosilylation of all active ≡Si-H group do not take place. During hydride addition the reaction order, reaction rate constants and activation energy were found.

IR and NMR spectra data determined the structures of synthesized oligomers. On the basis of NMR spectra data was offered, that the hydrosilylation proceeds mainly with formation intermediate product, by regrouping of this transition complex the oligomers with various structures of molecules are obtained. It was shown that at high lengths of side-substituted groups, the NMR spectra proved existence of conformers. Gel permeation chromatographic, differential scanning calorimetric and wide-angle X-ray analysis of synthesized oligomers was carried out.

Keywords: acrylates, methacrylate, hydrosilylation and methylhydridesiloxane.

---

* Corresponding author: mukbaniani@ictsu.tsu.edu.ge

INTRODUCTION

In literature, there are much information about hydrosilylation of acrylates and methacrylate [1-4]. By A.D. Petrov the reaction of hydride addition of triethylsilane to methyl acrylates in the presence of Spier catalyst has been investigated and it was shown that in hydrosilylation reaction 1,2- or 1,4-addition depend on the nature of organic group attached at silicon and on the structure of carbonyl compounds.

For example hydrosilylation of triethylsilane to methyl acrylates proceeds with 1,4-addition, while hydrosilylation of trichlorosilane to methyl acrylates proceeds with 1,2-addition. Consequently it was established, that hydrosilylation ability in triorganosilanes gradually changes with replacement of alkyl groups by electronegative halogens [1-3]:

$$(C_2H_5)_3SiH + CH_2{=}CH{-}C({=}O)OMe \xrightarrow{Cat} CH_3{-}CH{=}C(OSi(C_2H_5)_3)(OMe)$$

$$Cl_3SiH + CH_2{=}CH{-}COOCH_3 \longrightarrow Cl_3SiCH_2CH_2COOCH_3$$

during hydride addition of trichlorosilane to methacrylate obtaining α-isomer, may be explain that el-ectrophilic trichlorosilyl group attack to carbon atom which has more electronic density. In methyl acrylates electro donor methyl group in α-position, rises nucleophilic properties in C=C fragment of β-carbon atoms [4,5].

Hydrosilylation of methyldichlorosilane to methyl acrylates in the presence of platinum on the carbon proceeds by 1,2- and 1,4-addition [1-3]:

$$MeSiHCl_2 + CH_2{=}CH{-}COOMe \xrightarrow{Cat} \begin{cases} MeCl_2SiO{-}C(OMe){=}CH{-}CH_3 \\ MeCl_2Si{-}CH(CH_3){-}COOMe \\ MeCl_2SiCH_2CH_2COOMe \end{cases}$$

Hydrosilylation of triethylsilane to methyl methacrylate in the presence of catalyst Pt/C proceeds by 1,2-addition, with formation of triethyl(2-carbometoxypropyl)silane [1]:

$$Et_3SiH + H_2C{=}C(Me){-}C({=}O)OMe \xrightarrow{Cat} Et_3Si{-}CH_2{-}CH(Me){-}C({=}O)OMe$$

Hydrosilylation of methyldichlorosilane to acrylic esters in the presence of catalysts Pt/C or Pt/Al_2O_3 proceed with formation two isomeric products [6,7]:

$$CH_2{=}CHCOOR + MeSiHCl_2 \longrightarrow MeCl_2Si{-}CH(CH_3)COOR + MeCl_2Si{-}CH_2CH_2{-}COOR$$

In contrast to authors [1-7], hydrosilylation reactions of acrylates proceeds with formation of α, β and O-products, according to the following scheme [8]:

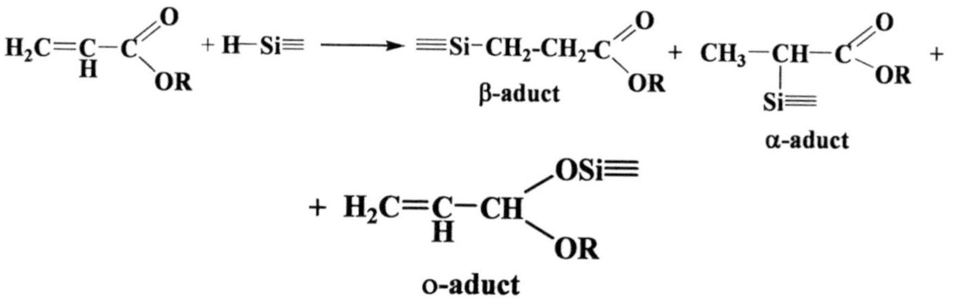

$$H_2C{=}\underset{H}{C}{-}C{\small\begin{matrix}{}^{\nearrow O}\\ {}_{\searrow OR}\end{matrix}} + H{-}Si{\equiv} \longrightarrow \underset{\beta\text{-aduct}}{\equiv Si{-}CH_2{-}CH_2{-}C(=O)OR} + \underset{\alpha\text{-aduct}}{CH_3{-}\underset{Si\equiv}{CH}{-}C(=O)OR} +$$

$$+\ \underset{\text{o-aduct}}{H_2C{=}\underset{H}{C}{-}CH(OSi{\equiv})(OR)}$$

For methyl acrylates the yield of β-addition product equal to -10,3 %, α-addition product – 57,8 % and o-addition product – 1,9 %. Substituted groups create a definite effect on formation of structure of hydrosilylation product.

The reaction of hydrosilylation of acrylate and methacrylate with 1,4-bis(dimethylsilyl)benzene in the presence of hydroquinone and Spier catalyst or platinum on the carbon was studied by the authors [9] and it was shown that the reaction proceeds with formation of corresponding acyloxysilanes, according to the following scheme:

$$2CH_2{=}CR{-}COOR' + H{-}Si(Me)_2{-}C_6H_4{-}Si(Me)_2{-}H \xrightarrow{Cat} R'OOC{-}C_2H_3R{-}Si(Me)_2{-}C_6H_4{-}Si(Me)_2{-}C_2H_3R{-}COOR'$$

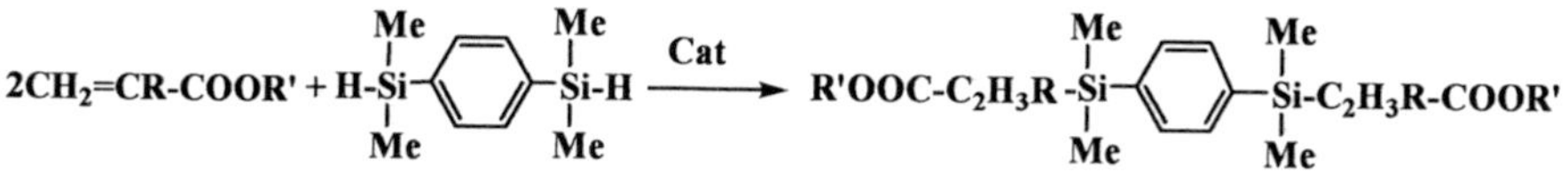

where: R=H, Me; R'=Me, Et.

In hydrosilylation reaction the hydroquinone was used for preventing the polymerization reactions.

The reaction of hydrosilylation of methylhydridesiloxane to unsaturated vinyl derivatives (alkenes, acrylic acid, acrylic esters) in the presence of Spier catalyst has been investigated by the authors [10] and methylsiloxane oligomers with carboxyl, alkyl and ester side-groups has been obtained, according to the following scheme:

$$\sim\!\underset{H}{\overset{Me}{Si}}{-}O{-}\underset{H}{\overset{Me}{Si}}{-}O{-}\underset{H}{\overset{Me}{Si}}{-}O\!\sim \; + n\, CH_2{=}CH{-}R \xrightarrow{H_2PtCl_6}$$

$$\sim\!\overset{Me}{Si}(CH_2CH_2R){-}O{-}\overset{Me}{Si}(CH_2CH_2R){-}O{-}\overset{Me}{Si}(CH_2CH_2R){-}O\!\sim$$

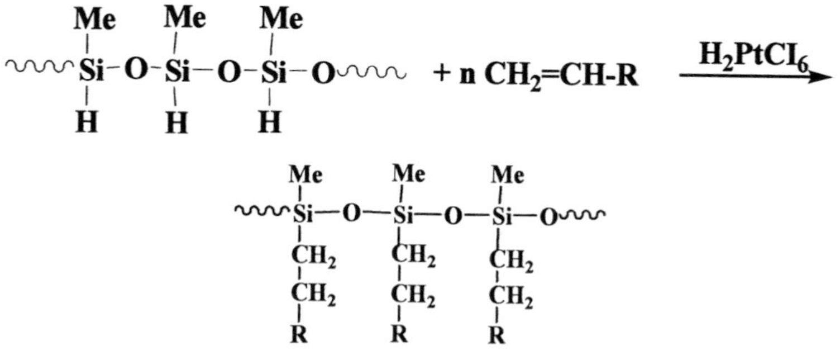

where: R=C_nH_{2n}, -COOH, -COOAlk.

Therefore, in literature there are contrasting information about course and mechanism of the hydrosi-lylation reaction of hydrideorganosilanes to acrylate and methacrylate.

EXPERIMENTAL PART

The starting materials for the synthesis of comb-type methylsiloxane oligomers were used α,ω–bis-(trimethylsiloxy)methylhydridesiloxane, acrylates and methacrylates.

The initial α,ω–bis(trimethylsiloxy)methylhydridesiloxane with the degree of polymerization n≈35, ac-rylate and methacrylate were received from Fluka. Before reaction the acrylates and methacrylate we-re distilled, or recrystallized. By drying and distillation cleaned the organic solvents used in reaction hydrosilylation.

IR spectra all of samples have been taken on an UR-20 instrument from KBr pellets, while the ^{1}H NMR spectra on a "Perkin-Elmer" at operating frequency of 250 MHz. All spectra were obtained with the use of $CDCl_3$ as solvent and internal standard. A Perkin – Elmer DSC-2 differential scanning calo-rimeter was used to determine TGA and the thermal transitions (T_g) were read at the maximum of the endothermic or exothermic peaks. Heating and cooling scanning rates were 10^0C/min.

Gel-permeation chromatography investigation was carried out with the use of Waters Model 6000A chromatograph with an R 401 differential refractometer detector. The column set comprised 10^3 and 10^4 Å Ultrastyragel columns. Sample concentration was approximately 3 % by weight in toluene and typical injection volume for the siloxane was 5 μL. Standardization of the GPC was accomplished by the use of styrene or polydimethylsiloxane standards with the known molecular weight.

Wide-angle X-ray diffractograms were taken on a "DRON-2" instrument. A-CuK_α was measured with-out a filter, the angular velocity of the motor ω≈2 deg/min.

Hydride Addition of α,ω–Bis(Trimethylsiloxy)Methylhydridesiloxane to Methyl Acrylates

The hydrosilylation reaction was carried out in a three-necked flask equipped with a mechanical stirrer, an argon inlet, reflux condenser and calcium chloride drying tube. The initial reagents were placed in the flask and thermostated in an oil bath until constant temperature was achieved. 0,1 M solution of Platinum hydrochloric acid in tetrahydrofuran ($5\div9x10^{-5}$ g per 1,0 g of starting substance) was used as a catalyst. The reaction carried out in an argon atmosphere in anhydrous toluene solution, at temperature range 60^0C. Then 5 ml anhydrous toluene was added to the reaction mixture, was filtered, connected to the vacuum and the toluene and unreacted methyl acrylates were removed at 30 $\div 40^0$C.

The hydrosilylation reaction in the presence of platinum on the carbon and with other acrylates and methacrylate has been carried out in the same manner.

RESULTS AND DISCUSSION

In literature there are known comb-type methylsiloxane polymers with various organic branching frag-ments containing mesogenic groups in the side chain [11-13]. These polymers characterized with li-quid-crystalline properties. Besides in literature there are known diorganosiloxane polymers with li-quid-crystalline properties not containing classical mesogenic groups. Such polymers corresponds me-thylpropyl-, diethyl- and dipropylsiloxane polymers with linear structure [14-16].

Present paper deal to synthesis and investigation of properties of comb-type methylsiloxane oligomers with various branching organic fragments. For the purpose of synthesis of siliconorganic comb-type oligomers with various ester groups in the side chain the reaction of hydride addition of α,ω–bis(tri-methylsiloxy)methylhydridesiloxane to acrylates and methacrylate in the presence of 0,1 M solution of Platinum hydrochloric acid (in tetrahydrofuran), or platinum on the carbon Pt/C has been investigated. The reaction carried out at 1:35 ratio of initial compounds, at various temperatures (30-80^{0}C), as in melt condition so in dilute solution of dry toluene.

During the hydride addition, the changes of concentration of active ≡Si-H group on the time were ob-served. The hydride addition was investigated in dry toluene solution (C≈5,3x10^{-2} mole/l). It was es-tablished, that during the reaction not all active ≡Si-H groups completely take place in hydride addi-tion. On the Figure 1 and 2 one can observe dependence of changes of concentration of active ≡Si-H groups on the time during of hydride addition of α,ω–bis(trimethylsiloxy)methylhydridesiloxane to methyl acrylate and methyl methacrylate.

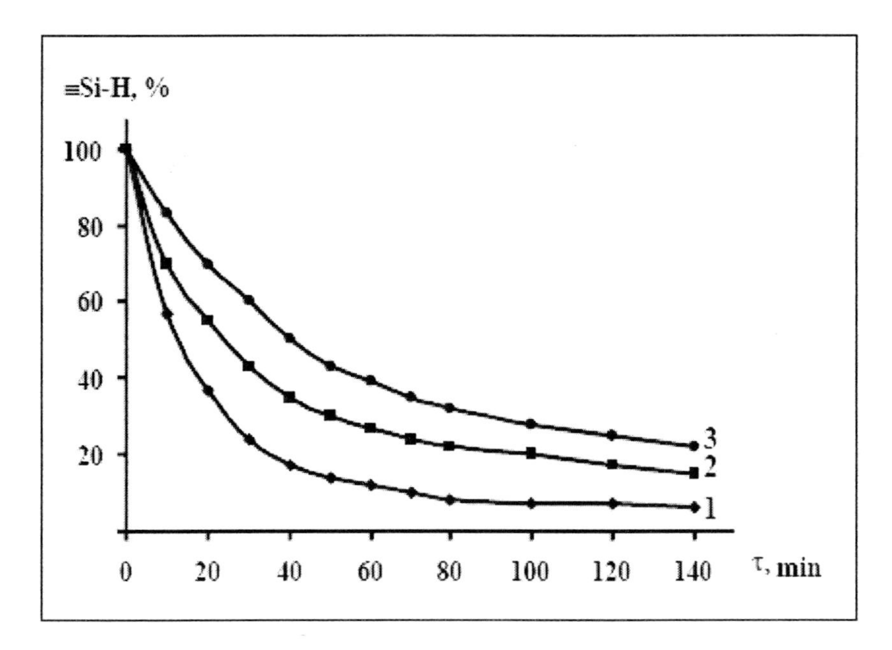

Figure 1. Dependence of changes of concentration of active ≡Si-H group on the time, upon hydride addition of α,ω–bis(trimethylsiloxy)methylhydridesiloxane to methyl acrylate, where curve 1 is at 50^{0}C temperature, curve 2 at 40^{0}C and curve 3 is at 30^{0}C.

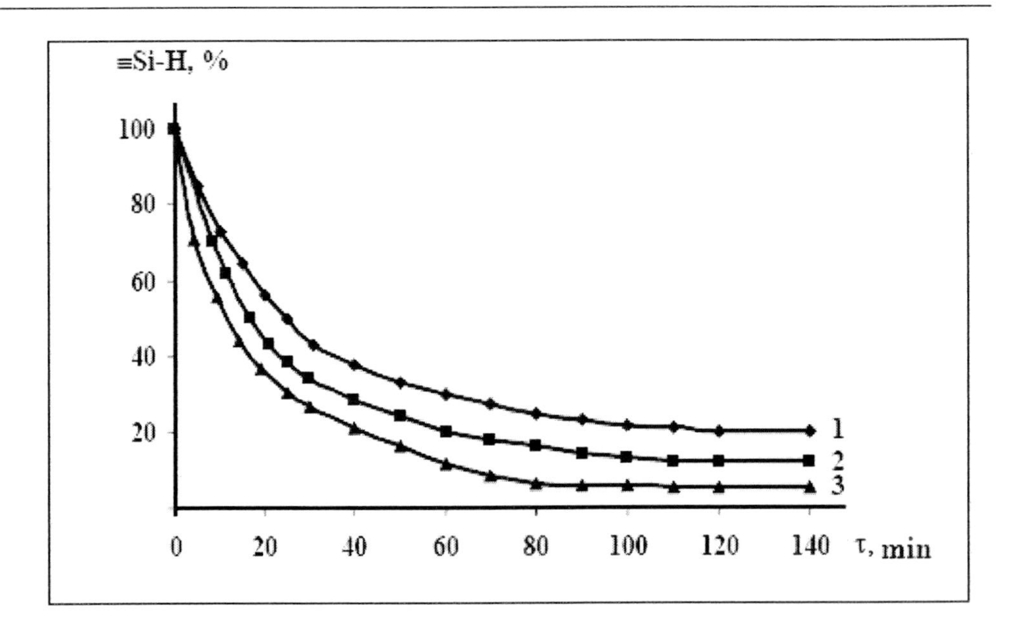

Figure 2. Dependence of changes of concentration of active ≡Si-H group on the time,upon hydride addition of α,ω–bis(trimethylsiloxy)methylhydridesiloxane to methyl methacrylate, where curve 1 is at 50^0C temperature, curve 2 at 55^0C and curve 3 is at 60^0C.

As it is seen from Figures 1 and 2 with an increase of temperature from 50^0C, to 60^0C the depth of hydride addition rises. At 50^0C temperature the depth of hydrosilylation in the time of methyl acrylate equal to 95% and in case of methyl methacrylate ~ 80%. Low rate of hydrosilylation in case of methyl methacrylate may be explained by steric effect produced by isomeric structure. With an increase of the length of unsaturated esters the rate and the depth of hydrosilylation decreases. In case of butyl me-thacrylate at 60^0C hydrogen conversion equal to ~78%.

So, hydride addition of methylhydridesiloxane to acrylate and methacrylate proceeds with formation various-linked oligomers according to the following scheme:

$$Me_3SiO-\left[\begin{matrix}Me\\|\\Si\\|\\H\end{matrix}-O\right]_m-SiMe_3 + mCH_2{=}CR-CO-OC_nH_{2n+1} \xrightarrow{H_2PtCl_6}$$

$$\longrightarrow Me_3SiO-\left\{\left[\begin{matrix}Me\\|\\Si\\|\\C_2H_3R\\|\\COOC_nH_{2n+1}\end{matrix}-O\right]_{(a)}-\left(\begin{matrix}Me\\|\\Si\\|\\H\end{matrix}-O\right)_{(b)}\right\}_{(c)}-SiMe_3$$

where: [(a)+(b)](c)=m≈35; R=H; n=1 [Me - 30^0C-I[1], 40^0C - I[2], 50^0C - I]; n=2 (C_2H_5 - II); n=4 (n-C_4H_9 -III); n=6 (n-C_6H_{13} - IV); n=16 (n-$C_{16}H_{33}$ - V).

m≈35; R=Me, n=1 [Me - 50^0C(VI[1]), 55^0C(VI[2]), 60^0C(VI)]; n=4 [C_4H_9 - 60^0C(VII[1]), 70^0C(VII[2]), 80^0C (VII)].

Hydride addition proceeds vigorously during first one hour, than the reaction rate decreases. The syn-thesized oligomers are vitreous liquid or solid products, depending on the surrounding group, well so-luble in aromatic type ordinary organic solvents with $\eta_{sp} \approx 0{,}03 \div 0{,}07$. The structure and composition of oligomers, proved by elementary analysis, determination of molecular masses, IR, ^{1}H and ^{13}C NMR spectral data. Some physical-chemical properties of synthesized oligomers are presented in Table 1 and 2.

Table 1. Some physical-chemical properties of synthesized comb-type oligomers.

#	Yield, %	R	η^*_{sp}	d_1, Å	T_g, ^0C	Elementary analysis, %**		
						C	H	Si
I^1	95	CH_3	0,05	8,63	-	*40,08* 39,79	*6,98* 6,37	*21,15* 20,87
I^2	89	CH_3	0,04	-	-	*39,34* 39,00	*6,62* 6,20	*22,44* 22,08
I	82	CH_3	0,03	-	-120	*38,51* 38,23	*6,98* 6,60	*23,26* 23,00
II	95	C_2H_5	0,08	10,10	-93	*44,50* 44,12	*7,61* 7,40	*18,63* 18,28
III	95	C_4H_9	0,05	-	-	*50,11* 49,79	*8,54* 8,29	*16,25* 15,68
IV	93	C_6H_{13}	0,03	15,37	-87	*54,17* 53,76	*9,23* 8,95	*14,57* 14,21
V	89	$C_{16}H_{33}$	0,05	27,1	-82	*66,02* 65,79	*11,13* 10,86	*9,14* 8,85

Table 2. Some physical-chemical properties of synthesized comb-type oligomers.

#	Yield, %	R	η^*_{sp}	d_1, Å	T_g, ^0C	Elementary analysis, %**		
						C	H	Si
1	2	3	4	5	6	7	8	9
VI^1	76	CH_3	0,04	8,23	-	*43,25* 42,87	*7,56* 7,21	*20,05* 19,87
VI^2	72	CH_3	0,04	-	-	*43,82* 43,24	*7,57* 7,31	*19,38* 19,07
1	2	3	4	5	6	7	8	9
VI	80	CH_3	0,06	7,23	-110	*44,51* 44,11	*7,59* 7,30	*18,55* 18,70
VII^1	73	C_4H_9	0,06	-	-	*51,38* 51,52	*8,84* 8,60	*16,21* 15,89
VII^2	75	C_4H_9	0,06	-	-	*51,83* 51,74	*8,87* 8,54	*15,75* 15,79
VII	70	C_4H_9	0,06	11,41	-88	*52,48* 52,32	*8,91* 8,87	*15,10* 14,78

*In 1% toluene solution, at 25^{0}C.

**Over line calculated values, under line found values.

In the IR spectra of synthesized oligomers, one can observe absorption band characteristic for asymmetric valence oscillation of ≡Si-O-Si≡ and ≡CO-O-C≡ bonds in the region 1020 and 1150 cm^{-1} respectively. Absorption bands of ≡Si-C≡ bonds at 1260 cm^{-1}, for valence oscillation of ≡C-H bonds at 2950-3100 cm^{-1} regions. It must be denote that with an increase of the length of substituted organic side- groups the absorption intensity in this range rises. In the spectra, one can observe absorption bands characteristic for unreacted ≡Si-H bonds at 2165 cm^{-1}.

For oligomers I-VII, ^{1}H and ^{13}C NMR investigation was carried out. As is seen from Figure 3, in ^{1}H NMR spectra of oligomer I, one can observe singlet signal for protons of methyl group in the fragment ≡SiMe and –$SiMe_3$ with chemical shift δ≈0,1 ppm, for protons of methylene groups in the fragment ≡Si-CH_2-, triplet signal with center of chemical shift δ≈0,85 ppm; for protons of methylene group, in the fragment –CH_2-CH_3 with chemical shift δ≈1,05 ppm. One can observe quartet signal for protons of methylene group in the fragment –O-CO-CH_2-, with chemical shift δ≈3,3 ppm, this signals in the fragment -CH_2-O-CO- undergo overlapping. Its displacement in low field is conditioned by influence of anisotropic cone of carbonyl group [17]. In spectra one can see low intensity singlet signal for protons of unreacted ≡Si-H group with chemical shift δ≈4,2 ppm.

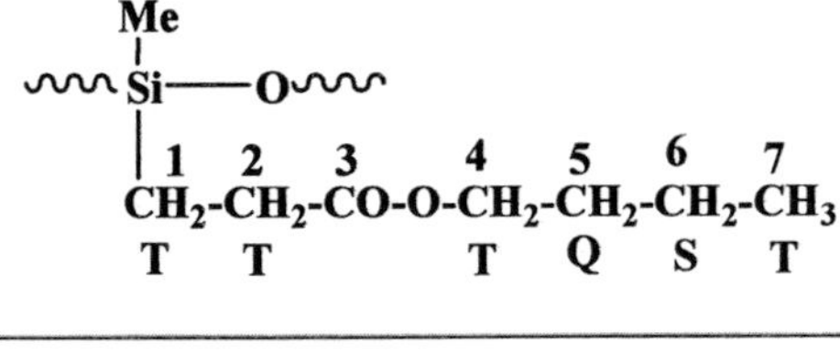

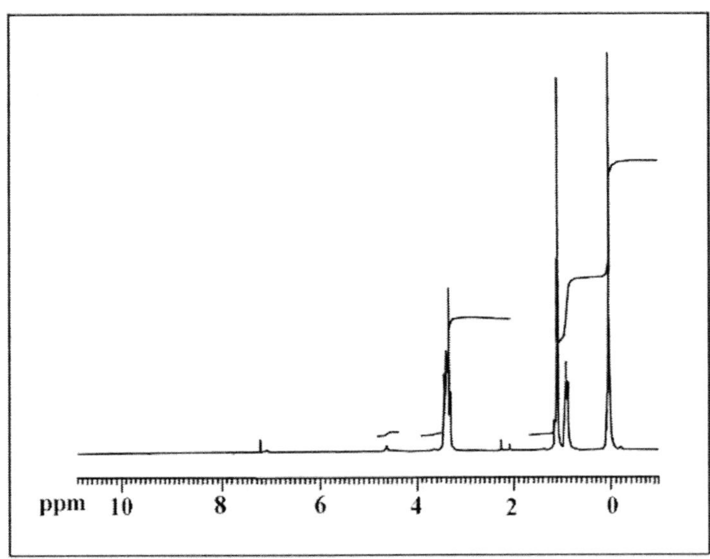

Figure 3. ^{1}H NMR spectrum of oligomer I.

In ^{1}H NMR spectra of oligomer II one can see singlet signals for methyl protons in the fragment –$SiMe_3$ and ≡Si-Me with center of chemical shifts δ≈0,1 ppm. In spectra one can see triplet signal cha-racteristic for methylene protons -C^1H_2-, -C^2H_2- and -C^4H_2- with center of chemical shift δ≈0,8; δ≈3,6 and δ≈3,95 ppm accordingly (Figure 4).

In ^1H NMR spectrum of oligomer II one can see quintet signals characteristic for protons of methylene group in fragment -C^5H_2- with the center of chemical shift δ≈1,55 ppm; sextet signals for protons of methylene group -C^6H_2- with the center of chemical shift δ≈1,20 ppm. It must be denote that quintet and sextet signal for protons of two methylene group -C^5H_2- and -C^6H_2- overlap to each other and resonate in the range δ≈1,55-1,20 ppm. As is seen from the spectrum probably, there are conformational isomers, which appear because of lengthy organic substituted side-groups. In ^1H NMR spectrum of oligomer III one can see triplet signals with center of chemical shift δ≈1,0 ppm, characteristic for protons of methyl -C^7H_3 groups. Starting from ^1H NMR spectrum of oligomers hydride addition reaction mainly proceeds by Farmer rule.

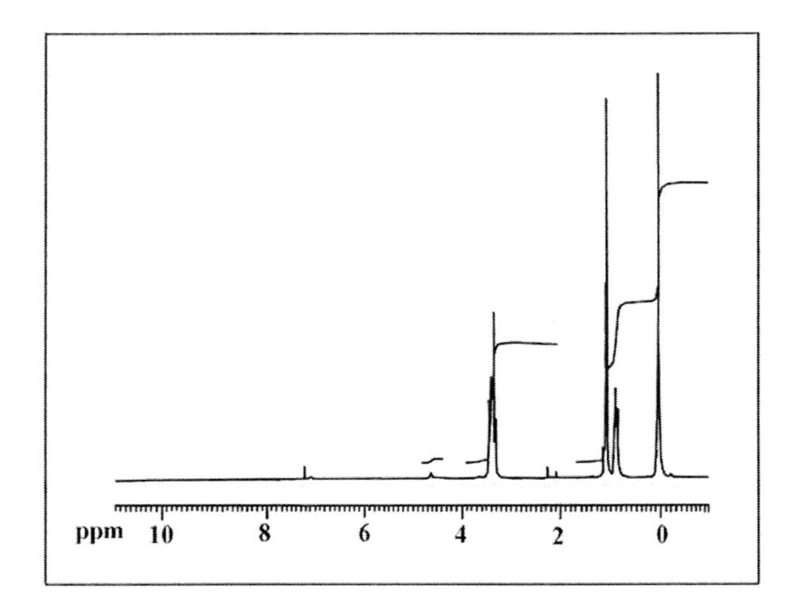

Figure 4. ^1H NMR spectrum of oligomer III.

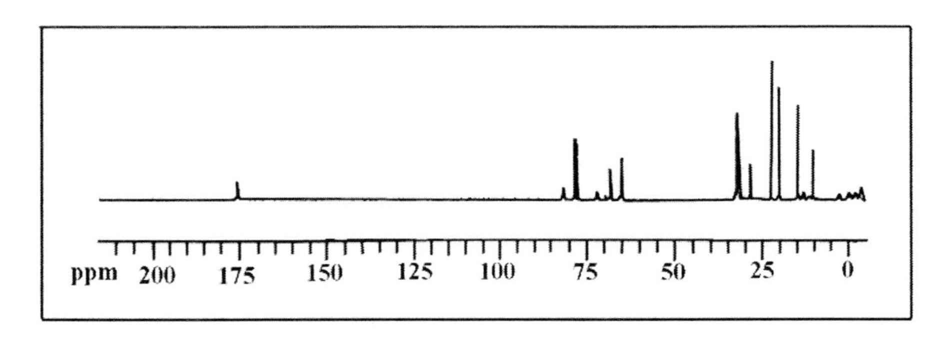

Figure 5. ^{13}C NMR spectrum of oligomer III.

Presence of two groups of conformational isomers was verified by ^{13}C NMR spectral data (Figure 5), where signals appear as a pair. For example, for ≡Si-Me groups one can see the signals with chemical shifts δ≈10,17 ppm and δ≈10,56 ppm, for -C^1H_2- groups one can see the signals with chemical shifts δ≈14,73 ppm and δ≈14,89 ppm; for -C^7H_3 groups signal with chemical shift δ≈22,41 ppm; for -C^5H_2- - δ≈28,60; for –C^4H_2- signals with chemical shift δ≈65,11 and 68,29 ppm. It must be denote that conformation most of all was observed at this fragment.

On the Figure 6 one can see ^1H NMR spectra of oligomer VI. From spectra it is evident that hydride addition of methylhydridesiloxane to methacrylate proceeds both by Farmer rule and by Markovnikov rule, by 1,2- and 1,4-addition (with the ratio 1:1,3). Therefore, based on spectral data of oligomers, it was proposed that in generally hydride addition of methylhydridesiloxane to methacrylate might be proceeds by several ways, see scheme 1:

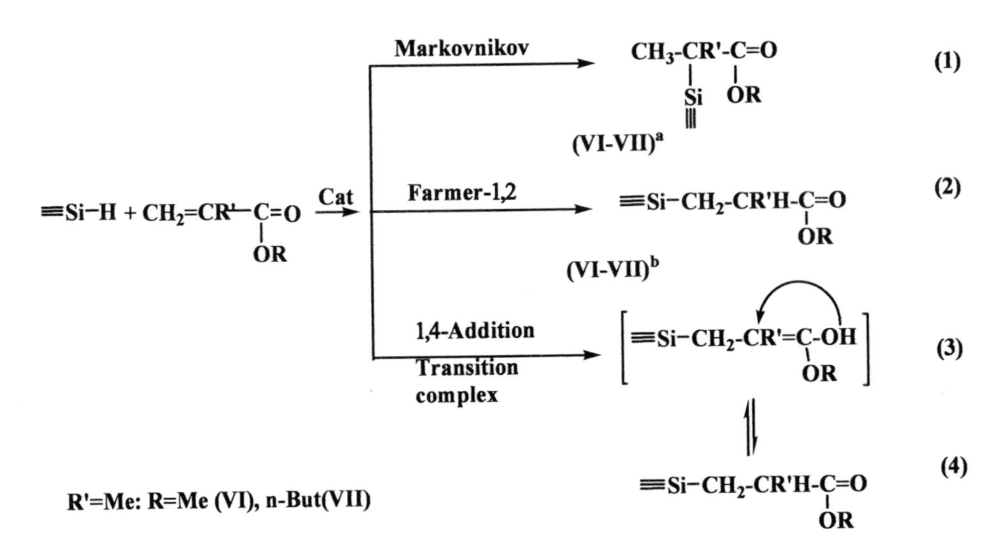

Scheme 1. The possible scheme of hydride addition of methylhydridesiloxane to acrylate and methacrylate.

As is seen from the possible scheme 1, in the case of 1,4-hydride addition reaction proceeds with ob-taining of intermediate-transition complex (3), by regrouping of this intermediate product according to the Eltekov rule (4), the product of 1,2-addition by Farmer rule are obtained. In ^1H NMR spectrum of oligomer VI (Figure 6) one can observe singlet signal characteristic for protons of methyl (Me-C≡, Markovnikov addition, scheme 1.1, structure VI^a) groups with chemical shift δ≈ 1,25 ppm, singlet signal of methyl protons (Me_3Si-) with chemical shift δ≈0,11 ppm; for methoxy groups (-COOMe) sharp singlet signal with chemical shift δ≈3,66 ppm.

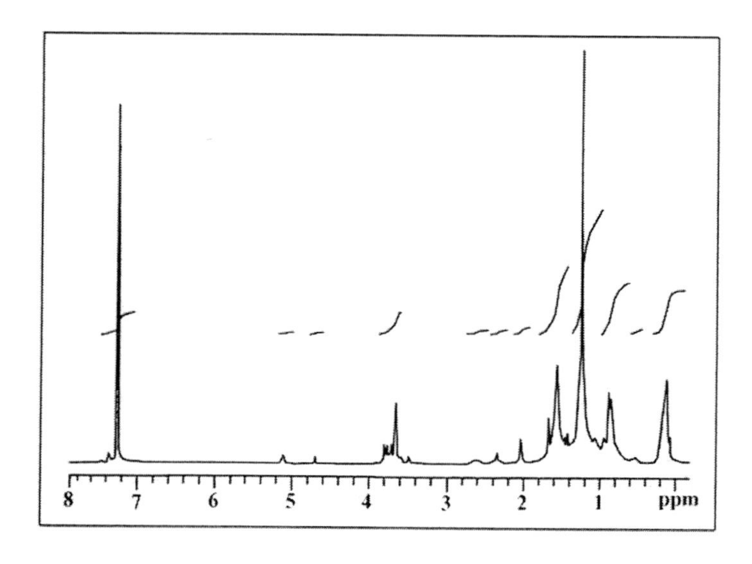

Figure 6. ^{1}H NMR spectrum of oligomer VI.

In ^{13}C NMR spectrum one can observe resonance signal for ester carbonyl group in weak field part, with chemical shift δ≈166,2 ppm; signal for methoxy groups (-COOMe) - δ≈41,4 ppm and resonance signal characteristic for quaternary carbon nucleus with chemical shift δ≈33,8 ppm; signal for methyl carbon (Me-CH) with chemical shift δ≈ 19,7 ppm (see Table 3 and 4).

Table 3. The chemical shifts of ^{1}H NMR spectrum of oligomer VI.

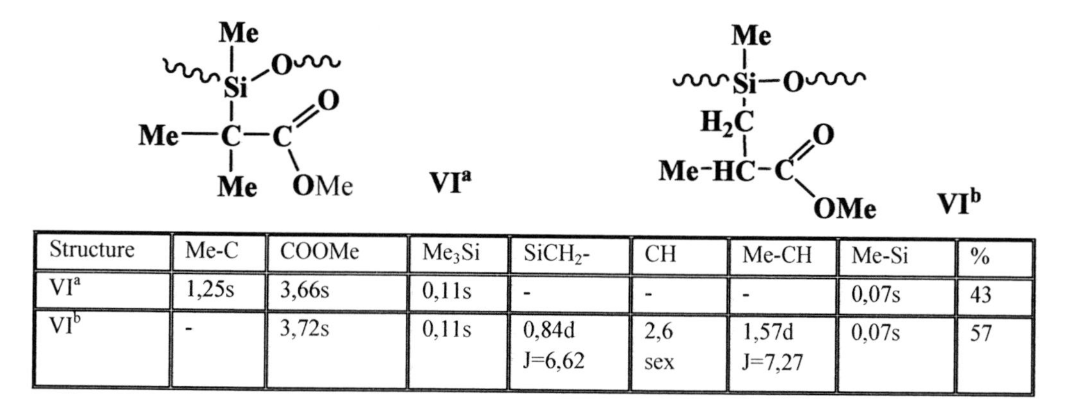

Structure	Me-C	COOMe	Me_3Si	$SiCH_2$-	CH	Me-CH	Me-Si	%
VI^{a}	1,25s	3,66s	0,11s	-	-	-	0,07s	43
VI^{b}	-	3,72s	0,11s	0,84d J=6,62	2,6 sex	1,57d J=7,27	0,07s	57

Table 4. The chemical shifts of ^{13}C NMR spectrum of oligomer VI.

Structure	Me-C	-C≡	OMe	C=O	$SiCH_2$-	CH	Me-CH	$SiMe_3$	≡SiMe
VI^{a}	19,7	33,8	41,4	166,2	-	-	-	14,2	1,8
VI^{b}	-	-	41,2	166,2	18,2	30,2	19,7	14,2	1,8

For oligomer obtained by Farmer rule (scheme 1.2 structure VI^{b}) ^{1}H NMR spectrum contain doublet signal for ≡Si-CH_2- groups with chemical shift δ≈0,84 ppm (J=6,62 Hz); sextet signal for protons of methine (≡CH) groups with chemical shift δ≈2,6 ppm; doublet signal for methyl protons (Me-CH) with chemical shift δ≈1,57 ppm and for methoxy groups resonance absorption at δ≈3,72 ppm. In ^{13}C NMR spectrum of oligomer VI (structure VI^b, Figure 7) one can see signal for ester carbonyl groups, which is analogically with structure VI^a resounds at δ≈166,2 ppm, additionally one can observe resonance signal for methine (≡CH) groups with chemical shift δ≈30,2 ppm and for methyl groups signal with chemical shift δ≈19,7 ppm. Signal with chemical shift δ≈18,2 ppm corresponds to methyl-lene groups in the fragment ≡Si-CH_2- (see Table 4).

For oligomer VII analogically of oligomer VI, ^1H and ^{13}C NMR spectral investigation has been carried out. The results of NMR investigations are presented in Tables 5 and 6.

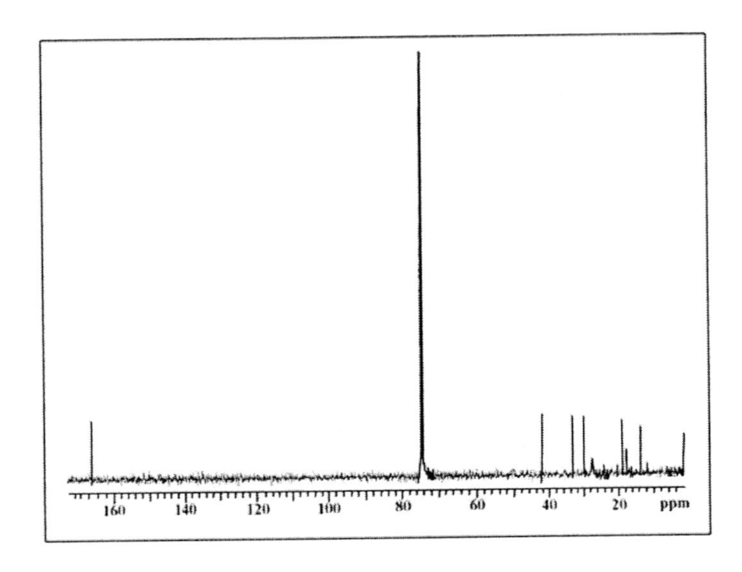

Figure 7. ^{13}C NMR spectrum of oligomer VI.

Table 5. The chemical shifts of ^1H NMR spectrum of oligomer VII.

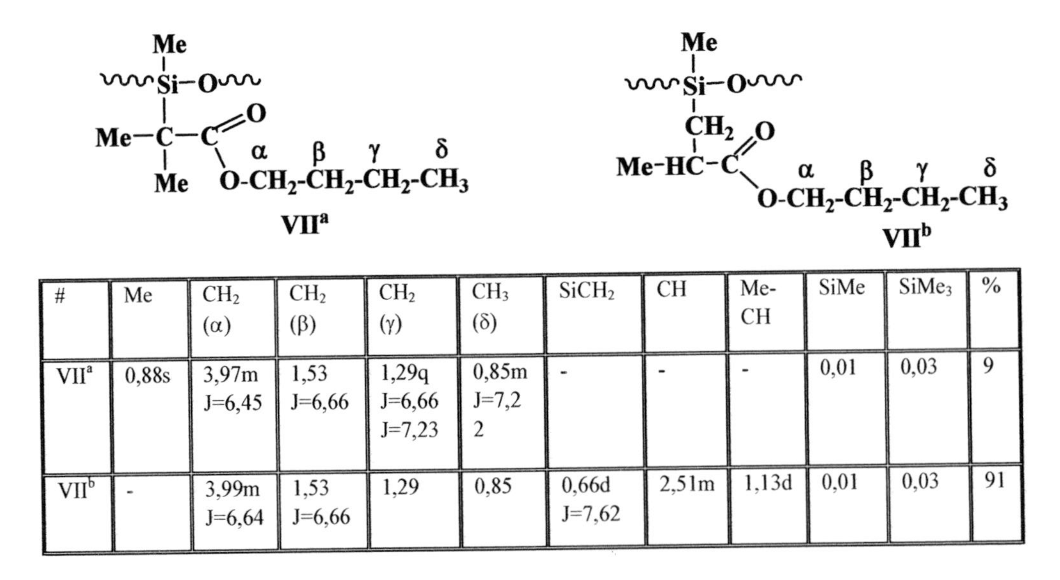

#	Me	CH_2 (α)	CH_2 (β)	CH_2 (γ)	CH_3 (δ)	$SiCH_2$	CH	Me-CH	SiMe	$SiMe_3$	%
VII^a	0,88s	3,97m J=6,45	1,53 J=6,66	1,29q J=6,66 J=7,23	0,85m J=7,2 2	-	-	-	0,01	0,03	9
VII^b	-	3,99m J=6,64	1,53 J=6,66	1,29	0,85	0,66d J=7,62	2,51m	1,13d	0,01	0,03	91

Table 6. The chemical shifts of ^{13}C NMR spectrum of oligomer VII.

#	Me	CH_2 (α)	CH_2 (β)	CH_2 (γ)	CH_3 (δ)	$SiCH_2$	CH	Me-CH	COO	SiMe	$SiMe_3$	-C≡
VII[a]	19,0	64,6	19,4	21,9	14,1	-	-	-	177,1	1,9	2,03	34,8
VII[b]	-	64,8	19,4	21,9	14,1	20,1	30,8	19,3	177,4	1,9	2,03	-

For fully characterization of hydride addition of methylhydridesiloxane to methyl methacrylate by quantum-chemical half empiric AM1 method [18], for all initial, intermediate and final products, in model reaction of hydrosilylation of methyldimethoxysilane to methyl methacrylate, it was calculated heats of formations (ΔH_f) and heat of reaction ($\Delta\Delta H$) proceeding with various concurrent scheme 2:

MeSi(OMe)$_2$-H + $H_2C=C(Me)-C(=O)OMe$ —Cat→

Markovnikov → Me–C(Me)(Si(Me)(OMe)MeO)–C(=O)OMe (1)
VI[a] $\Delta\Delta H_1$= -74,57Kj/mole

Farmer-1,2 → Me-Si(OMe)$_2$-CH_2-CH(Me)-C(=O)OMe (2)
VI[b] $\Delta\Delta H_2$= -137,44Kj/mole

1,4-Addition Transition Complex → [Me-Si(OMe)$_2$-CH_2-C(Me)=C(OMe)-OH] (3)
$\Delta\Delta H_3$ = -74,15 Kj/mole

2.3.a. [Me-Si(OMe)$_2$-CH_2-C(Me)=C(OMe)-OH] → Me-Si(OMe)$_2$-CH_2-CH(Me)-C(=O)OMe
$\Delta\Delta H$= -74,15 Kj/mole $\Delta\Delta H_2$= -63,29 Kj/mole

Scheme 2. The possible scheme of hydride addition of methylhydridesiloxane to methyl methacrylate.

As is seen from the scheme 2, starting from heat of reaction of hydrosilylation of methyldimetho-xysilane to methyl methacrylate, thermodynamically is more profitable the course of reaction according to the Farmer rule (1,2-addition). It must be denote that the product obtained by regrouping of inter-mediate product (1,4-addition, scheme 2.3) similar to the Eltekov rule (scheme 2.3a), identical to struc-ture of the product obtained by Farmer rule (scheme 2.2) and they have the same heat of reaction.

On the Figure 8 dependence of reverse concentration of the reactant products on the time is presented. Figure 8 shows dependence of reverse concentration on the time during hydride addition of methyl-hydridesiloxane to methacrylate. One can see that at the initial stages the hydrosilylation reaction is of second order. The hydrosilylation rate constants at various temperatures were calculated. For methacry-late: K_{30^0C}≈0,3838, K_{40^0C}≈0,7813 and

$K_{50^0C} \approx 1{,}6003$ mole/l·s ($\gamma \approx 2$); for methyl methacrylate: $K_{50^0C} \approx 1{,}9314$, $K_{55^0C} \approx 3{,}3286$ and $K_{60^0C} \approx 4{,}1006$ mole/l·s; for butyl methacrylate: $K_{60^0C} \approx 0{,}9754$, $K_{70^0C} \approx 1{,}5072$ and $K_{80^0C} \approx 3{,}3286$ mole/l·s ($\gamma \approx 2$).

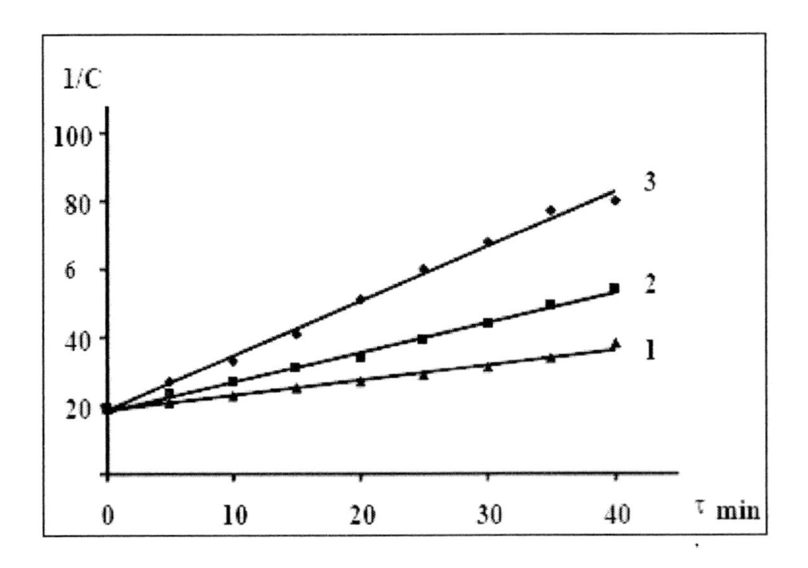

Figure 8. Dependence of the reverse concentration on the time, upon hydride addition of methylhydridesiloxane to methacrylate. Where curve 1 is at 30^{0}C, curve 2 is at 40^{0}C and curve 3 is at 50^{0}C.

Figure 9 shows the dependence of hydrosilylation rate constants logarithm of the on the inverse temperature upon of hydride addition of methylhydridesiloxane to methyl acrylate. From these data the activation energy of hydrosilylation reaction was calculated. For methyl acrylate, it is equal to: $E_{act} \approx$ 55,7 kJ/Mole; for methyl methacrylate - $E_{act} \approx 58{,}0$ kJ/Mole and for butyl methacrylate - $E_{act} \approx 62{,}1$ kJ/Mole. As is seen from above-mentioned values, the activation energies a little distinguished from each other.

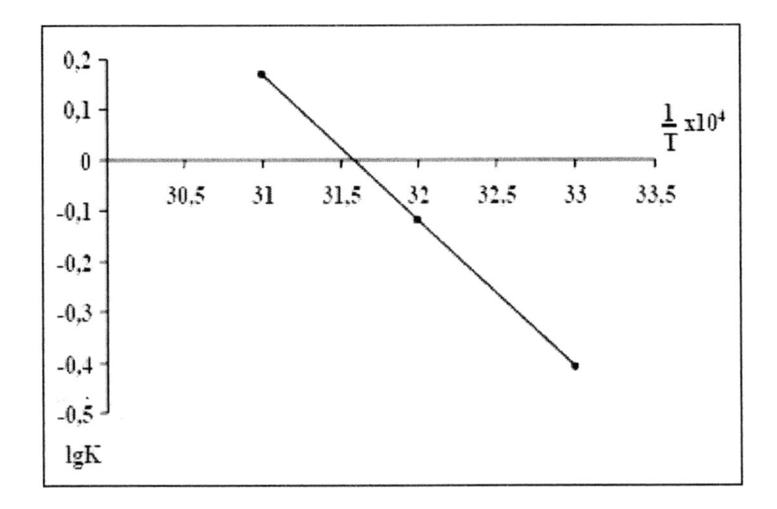

Figure 9. Dependence of hydrosilylation reaction rate constants logarithm of the on the in-verse temperature during hydride addition of methylhydridesiloxane to methyl acrylate.

The synthesized oligomers II-IV were studied by gel permeation chromatographic method. Figure 9-11 shows the molecular we-ight distribution of oligomers II, IV and V. From this figures, it is evident that the oligomers have bi- and trimodal molecular weight distribution. The molecular masses of oligomers were determined.

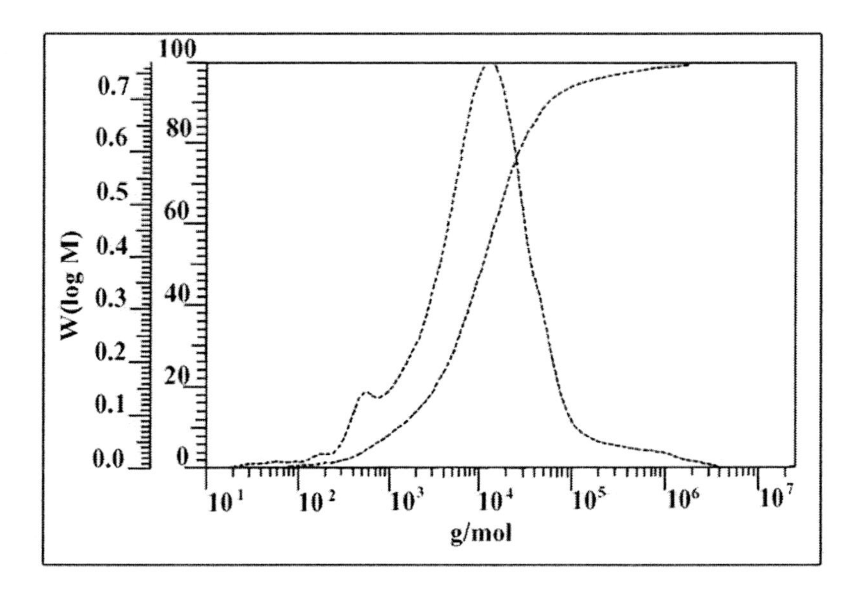

Figure 9. Gel permeation chromatographic curve of oligomer II.

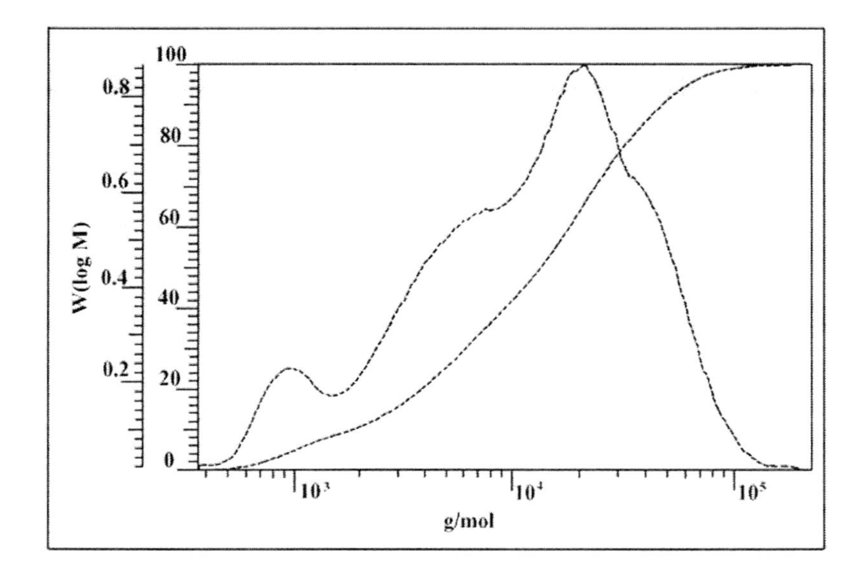

Figure 10. Gel permeation chromatographic curve of oligomer IV.

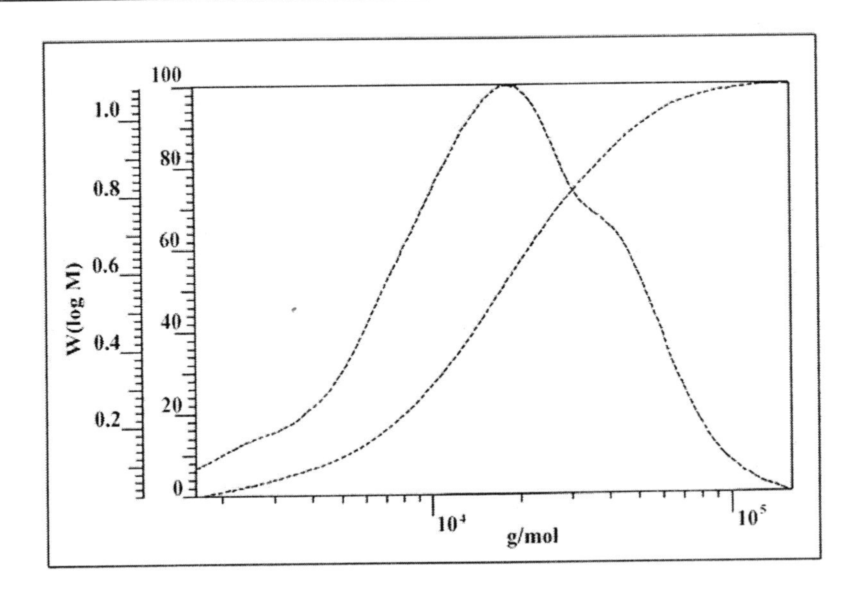

Figure 10. Gel permeation chromatographic curve of oligomer V.

The average molecular mass of oligomers II is ~ $\overline{M}_\omega\approx56{,}66x10^3$, $\overline{M}_n\approx1{,}877x10^3$, $\overline{M}_z\approx2{,}6661x10^3$ and polydispersity D≈30. As it is seen from the values of average weight molecular masses it 11,4 times exceeds to theoretical calculated molecular mass (in case of full hydrosilylation). Analogically were determined average molecular masses for oligomers IV: $\overline{M}_\omega\approx4{,}587x10^4$, $\overline{M}_n\approx1{,}125x10^4$, $\overline{M}_z\approx4{,}193x10^4$, D≈4,08; for oligomers V: $\overline{M}_\omega\approx4{,}901x10^4$, $\overline{M}_n\approx2{,}00x10^4$, $\overline{M}_z\approx4{,}219x10^4$, D≈2,09. Increas-ed molecular masses proved that partially the hydrosilylation reaction proceeds intermolecular on the carbonyl group, which is proved with literature data [1]. It must be denote that in the work [1] was not used NMR investigation, which uniquely will be proved the structure of obtained products.

So, on the basis of NMR spectra and GPC data it is possible to prove that hydrosilylation mainly proceeds by 1,2-addition and partially proceeds intermolecular hydride addition of residual ≡Si-H gro-ups on carbonyl group. Though, in NMR spectrum characteristic resonance signals for =CH-O-Si≡ bonds does not observed, which may be explained with low technical possibility of used spectrometer.

By wide-angle X-ray investigation Figure 11, it was shown, that synthesized oligomers are one-phase amorphous systems. With an increase of the length of substituted side-groups in oligomers, the value of interchain distances rises from d_1=8,23 Å (oligomer VI) up to d_1=27,1 Å (oligomer V). The values of interchain distances of above-mentioned oligomers are near to value of interchain distances analogica-lly structure organic polymers [19].

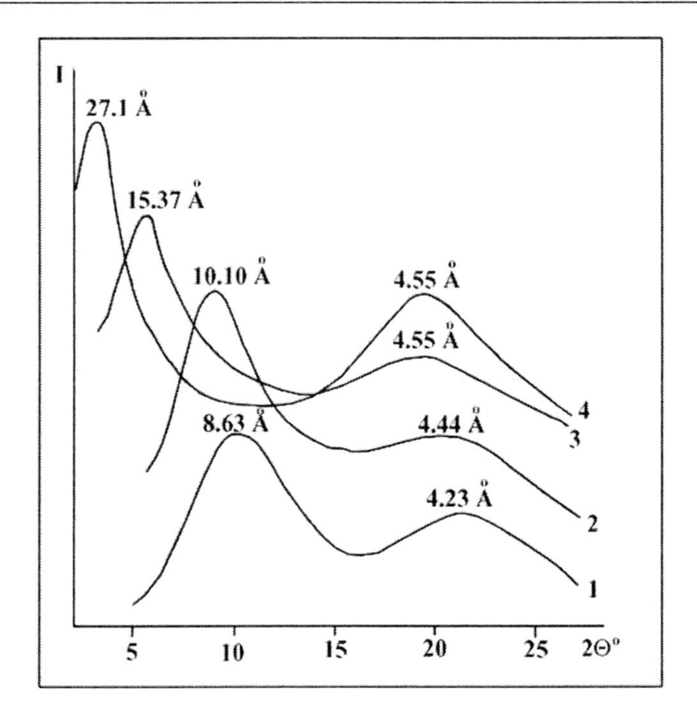

Figure 11. Wide angle X-ray diffraction patterns of oligomers, where curve 1 1 corresponds to oligomer I, curve 2 - II, curve 3 –IV and curve 4 – V.

By differential scanning calorimetric investigation it was shown, that in oligomers II and IV (Figure 12) one can observe one temperature transition in the range -82÷-93^{0}C, which corresponds to glass transition temperature of oligomers. On calorimetric curve of oligomer V (Figure 13) one can see temperature transition at +24,9^{0}C.

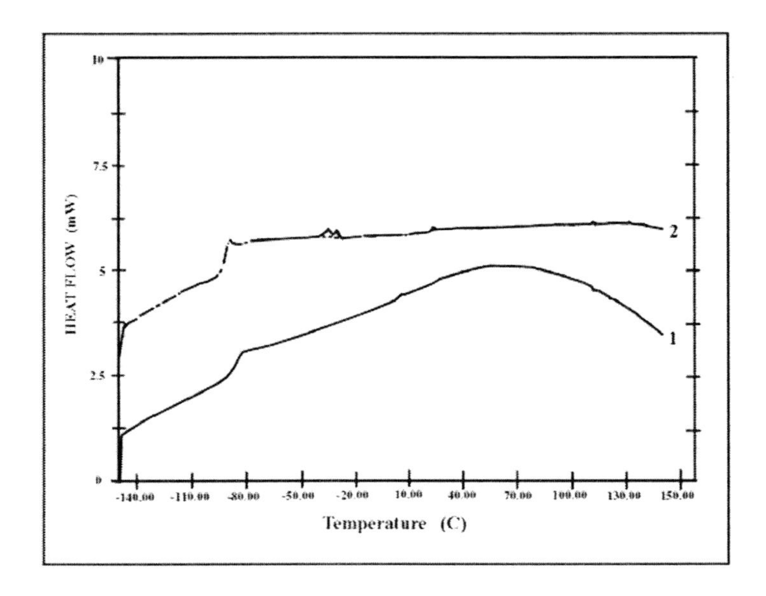

Figure 12. Differential scanning calorimetric curves of oligomers, where curve 1 corresponds to oligomer IV and curve 2 – II.

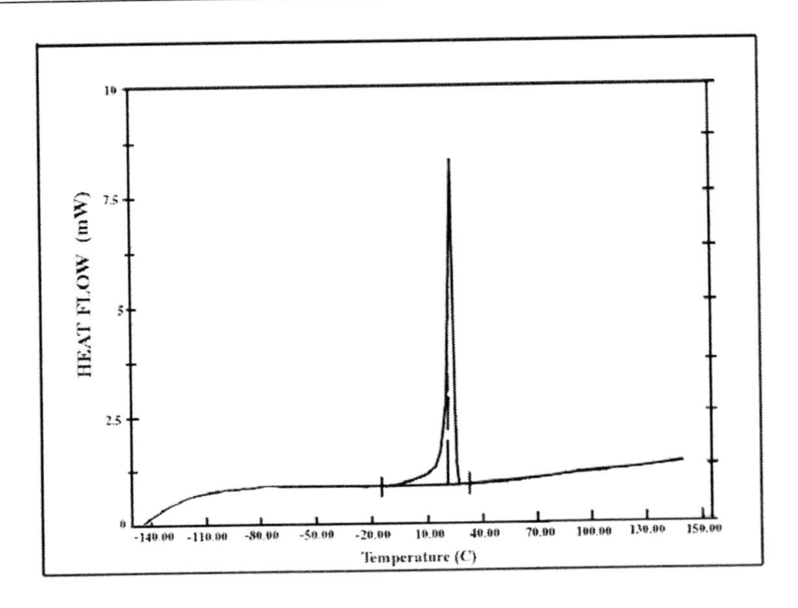

Figure 13. Differential scanning calorimetric curve of oligomer V.

Therefore, with an increase of the length of substituted side-groups the glass transition temperature rises. As is seen from DSC curves of oligomers, they do not reveal liquid-crystalline properties, in contrast to methylsiloxane oligomers containing various length oxyalkyl side groups [20].

So, by us for the first time the reaction of hydride addition of methylhydridesiloxane to acrylates and methacrylate has been studied and synthesized comb-type methylsiloxane oligomers with various length ester side groups. They are interesting products, because by its hydrolysis is possible to obtain water-soluble siliconorganic acids and their corresponding salts.

References

[1] Petrov A.D., Sadikh-Zade S.I., and Filatova E.I. , (1959). *Zhurn. Obshch. Khim.*,.29, 2936.

[2] Sadikh-Zade S.I., and Petrov A.D. (1958). *Chemistry and practical application siliconorganic compounds*, part 1, Leningrad, 1, 212.

[3] Sadikh-Zade S.I., and Petrov A.D. (1959). *Zhurn., Obshch. Khim.*, 29(10), 3194.

[4] Sommer L.M., Makray F.P., Stevard D.W., and Gampbeh P.G. (1957). *J. Amer. Chem. Soc.*, 79, 2764.

[5] Curry J.W., and Harrison C.W. (1958). *J. Org. Chem.*, 23, 627.

[6] Spier J.I., Webster J.A., and Bernes G.H. (1957). *J. Amer. chem., Soc.*, 79, 974.

[7] B.P. 882101 (1961). (Union Carbide Corpor.). "Chlorosilanes and derivatives from Acrylates". Pike R.A., Black W.Th.

[8] Chengyou Kan, Xiangzheng Kong, and Deshan Liu. (2000). *Polymer Preprints,* 41(2), 1243.

[9] Gverdciteli I.M., Doksopulo T.P., and Chikovani E.I. (1975). *Zhurnal Obshch. Khim.*, 47(7), 352.

[10] Nitzsche S. (1960). *Angew. Chem.*, 72(3), 112.

[11] Finkelmann H., Kock Y.J., and Rehage G. (1981). *Macromol. Chem., Rapid Commun.* 2, 317.

[12] Rinsdorf H., and Schneller A. Macromol. (1982). *Chem., Rapid Commun.*, 2, 557.

[13] Kawatsuki N., Sakashita S., Takatani K., Yamomoto T., and Sangen O. (1996). *Macromol. Chem., Phys.*, 197, 1919.

[14] Godovsky Yu.K., Mamaeva I.I., Makarova N.N., Papkov V.S., and Kuzmin N.N. (1985). *Macromol. Chem., Rapid Commun.*, 6, 797 .

[15] Godovsky Yu.K., Makarova N.N., Papkov V.S., and Kuzmin N.N. (1985). *Vysokomol., Soed.*, 27B, 164.

[16] Blumshtein A. (1981). "Liquid Crystalline Order in Polymers", *edited by Academic Press, New York* 1978, *translated into Russian, Mir, Moscow* 352.

[17] Hesse M., Meier H., and Zeeh B. (1979). "Spektroskopische Methoden in der Organischen Chemie". Georg Thieme Verlag, Stuttgart 480.

[18] Dewar M.I.S., Zeobish E.G., Healy E.F., and Stewart J.S. (1985). *J. Am. Chem. Soc.*, 107, 3902.

[19] Plate N.A., and Shibaev V.P. (1980). "Comb-type polymers and liquid crystals"; *Moscow, "Khimia"*, 17.

[20] Gurgenidze G.N. (2003). Candidate dissertation, Tbilisi State University, Tbilisi Georgia.

In: Reactions and Properties of Monomers and Polymers ISBN: 1-60021-415-0
Editors: A. D'Amore and G. Zaikov, pp. 21-41

Chapter 2

SOLUTION OF THE PROBLEM OF SELECTIVE OXIDATION OF ALKYLARENS BY MOLECULAR OXYGEN INTO CORRESPONDING HYDROPEROXIDES. CATALYSIS INITIATED BY NI(II), CO(II) AND FE(III) COMPLEXES ACTIVATED BY ADDITIVES OF ELECTRON-DONOR MONO- OR MULTIDENTATE EXTRA-KIGANDS

L. I. Matienko
N. M. Emanuel's Institute of Biochemical Physics Russian Academy of Sciences
4, Kosygin str., Moscow 111999, Russia

ABSTRACT

Mechanism of discovered by author and his collaborators [21-26] phenomenon of activity increase of complexes $M(L^1)_2$ (M=Ni(II), Co(II), L^1=acac$^-$) as catalysts of alkylarens (ethylbenzene, cumene, toluene) oxidation by molecular O_2 in ROOH in the presence of electron-donor additives of monodentate ligands L^2 (L^2 = HMPA (hexamethylphosphorus triamide), dimethyl formamide (DMFA), N-methyl pyrrolidone-2 (MP)), MSt (M = Li, Na, K) was studied. On the base of established mechanism of $Ni(II)(L^1)_2$ complexes catalytic activity control by monodentate ligands-modifiers L^2 in ethylbenzene oxidation reaction the optimization methods of processes of selective ethylbenzene oxidation into α-phenylethylhydroperoxide were proposed and realized. Values of selectivity, conversion degree and ROOH yield reached at that exceed analogous parameters in the presence of $\{Ni(II)(L^1)_2+L^2\}$ and known catalysts of ethylbenzene oxidation into ROOH. The method of estimation of selective catalytic particles activity formed in the course of ethylbenzene oxidation and at elementary process stages is suggested.

Keywords: oxidation, ethylbenzene, α-phenylethylhydroperoxide, homogeneous catalysis, molecular oxygen, nickel (II) bis(acetylacetonate), nickel (II) bis(enaminoacetonate), cobalt (II) bis(acetylacetonate), iron (III) tris(acetylacetonate), macrocycle polyether 18K6, quaternary ammonium salts R_4NBr.

INTRODUCTION

Presented Chapter is devoted to studies being the continuation and development of achievements in the field of homogeneous catalysis of organic substances oxidation by transition metals which are reflected in works of N.M. Emanuel's school [1-3], including author's works also [4-6].

Works of N.M. Emanuel and his school in the field of homogeneous catalysis are basic ones, since in them they established for the first time that transition metals compounds participated in all elementary stages of chain oxidation process [1-6]. Later on these discoveries were confirmed and described in reviews and monographs [7-13]. However, there is no complete understanding of mechanism yet. Special attention was attended to investigation of role of metals compounds at stages of free radicals generation, in chain initiation reactions (activation by O_2) and hydroperoxides dissociation. Reaction of chain propagation under interaction of catalyst with peroxide radicals (Cat + RO_2 →) is studied insufficiently. Catalysis by nickel compounds was studied in details only in works of author together with Z.K. Maizus, L.A. Mosolova (Goldina), I.P. Skibida, E.M. Brin [6].

They proved by kinetic and physical methods that formation of oxidation products occurred via formation of intermediate complexes of catalysts with active particles: Cat–O_2, Cat–ROOH, Cat–RO_2 [14]. Application of transition metals salts rarely leads to significant rise of process selectivity, since transformations of all intermediate substances are accelerated not selectively [14]. Not numerous examples of selective radical-chain oxidation of alkylarens with the use of catalytic systems on the base ob transition metals compounds are known. Mainly these are the processes leading to formation of products of deep oxidation. One of the most striking examples is oxidation of alkylarens into carbonyl compounds and carbonic acids by oxygen in the presence of so-called MC-catalysts (Co (II) and Mn (II) acetates, HBr, HOAc) [15].

Application of metal-complex catalysis opens possibility of regulation of relative rates of elementary stages Cat–O_2, Cat–ROOH, Cat–RO_2 and in that way to control rates and selectivity of processes of radical-chain oxidation [14]. Besides, initial catalyst form is often only the precursor of true catalytic particles and functioning of catalyst is always accompanied by processes of its deactivation. Introduction into reaction of various ligands-modifiers may accelerate formation of catalyst active forms and prevent or trig processes leading to its deactivation.

Solution of problem of selective oxidation of hydrocarbons into hydroperoxides, primary products of oxidation is the most difficult one. High catalytic activity of the majority of used catalysts in ROOH decomposition doesn't allow suggesting of selective catalysts of oxidation into ROOH to present day.

At recent decades the interest to fermentative catalysis and investigation of possibility of modelling of biological systems able to carry out selective introduction of oxygen atoms by

C–H bond of organic molecules (mono- and dioxygenase) is grown. Unfortunately, dioxygenases able to realize chemical reactions of alkanes dioxygeneration are unknown [16].

For alkylarens, hydrocarbons with activated C–H bonds (cumene, ethylbenzene) the problem of oxidation into ROOH at conditions of radical-chain oxidation process with degenerate branching of chain is solvable, since selectivity of oxidation into ROOH at not deep stages (~1-2%) is high enough (S ~ 80-90%). In this case the problem is in increase of reaction rate and conversion of hydrocarbon transformation into ROOH at maintaining of maximum reachable selectivity. Obviously, effective catalysts of oxidation into ROOH should possess activity in relation to chain initiation reactions (activation by O_2) accelerating formation of ROOH and also should be low effective in reactions of radical decomposition of formed during oxidation process active intermediates [14]. In spite of theoretical interest the problem of selective oxidation of alkylarens (ethylbenzene and cumene) in ROOH is of current importance from practical point of view in connection with ROOH use in large-tonnage productions such as production of propylene and styrene (α-phenylethylhydroperoxide), or phenol and acetone (cumyl hydroperoxide) [17].

Except catalytic systems developed by us nobody had proposed effective catalysts of selective oxidation of ethylbenzene into α-phenylethylhydroperoxide (PEH) until present days in spite of the fact that ethylbenzene oxidation process was well studied and a large number of publications and books were devoted to it [14, 18]. For example in the presence of homogeneous and heterogeneous catalysts on the base of transition metals compounds the selectivity of ethylbenzene oxidation into PEH is $S \leq 90\%$ at conversion degree (by spent RH) equal to $C \leq 5\%$ [18-20].

The method of homogeneous catalysts modifying by additives of electron-donor mono- and multidentate ligands for increasing of selectivity of liquid-phase oxidation processes was proposed by us.

For the first time we found the phenomenon of significant rise of initial rate (w^0), selectivity ($S = [PEH] / \Delta[RH] \cdot 100\%$) and conversion degree ($C = \Delta[RH] / [RH]_0 \cdot 100\%$) of oxidation of alkylarens (ethylbenzene, cumene, toluene) into ROOH by molecular O_2 under catalysis by transition metals complexes $M(L^1)_2$ (M = Ni(II), Co(II), L^1=acac$^-$) in the presence of additives of electron-donor monodentate ligands (L^2 = HMPA (hexamethylphosphorus triamide), dimethyl formamide (DMFA), N-methyl pyrrolidone-2 (MP)), MSt (M = Li, Na, K) [21-26]. Selectivity and conversion degree of hydrocarbon into hydroperoxide depend on hydrocarbon structure and may be regulated by selection of corresponding catalyst concentrations and ratio of system's components [26].

On the example of ethylbenzene oxidation (120°C) the mechanism of control of $M(L^1)_2$ complexes catalytic activity by additives of electron-donor monodentate ligands L^2 (L^2 = HMPA, DMFA, MP, MSt) was established.

Two catalytically active complexes are formed in the course of oxidation initiated by L^2. At initial stages of reaction the coordination of L^2 with $M(L^1)_2$ inhibits processes leading to decrease of catalyst activity and prevents the formation of complexes with phenol, one of the products of ethylbenzene oxidation catalyzed by $M(L^1)_2$, $M(L^1)_2 \cdot PhOH$ which carry out heterolysis of α-phenylethylhydroperoxide (PEH) and possess ability to inhibit oxidation process [27, 28]. Oxidation-reduction activity of formed in situ primary complexes $M(L^1)_2L^1$ is increased that is expressed in the rise of rate of chain initiation (activation by O_2) and homolytic decomposition of PEH [21, 23]. In this connection selectivity of oxidation into

PEH at initial stages of reaction is not high. However initial rate of reaction is increased. With process development the increase of S_{PEH} ($S_{PEH}^{max} \approx 90\%$) in comparison with initial stage ($S_{PEH}^{max} = 80\%$) of oxidation and decrease of reaction **w** are observed. In developed reaction ligands L^2 control transformation of $M(L^1)_2$ complexes into more active selective particles. At that the rise of S_{PEH} is reached at the expense of catalyst participation in activation reaction of O_2 and inhibition of chain and heterolytic decomposition of PEH. Direction of side products acetophenone (AP) and methylphenylcarbinol (MPC) formation is changed from consequent (under hydroperoxide decomposition) to parallel [30].

The mechanism of $M(L^1)_nL^2$ complexes transformation depends on metal ion (M=Ni(II), Fe(III), Co(II), L^1=acac$^-$) and complex's chelate group (M=Ni(II), L^1=acac$^-$, enamac$^-$ (enaminoacetonate)). We established that in the case of nickel complexes selective catalyst was formed as result of controlled by L^2 ligand regio-selective connection of O_2 to nucleophilic carbon γ-atom of one of the ligands L^1. Coordination of electron-donor extra-ligand L^2 by nickel complex $Ni(L^1)_2$ (L^1=acac$^-$) promoting stabilization of intermediate zwitter-ion $L^2(L^1M(L^1)^+O_2^-)$ leads to increase of possibility of regio-selective connection of O_2 to acetylacetonate ligand activated in complex with nickel (II) ion. Further introduction of O_2 into chelate cycle accompanying by proton transfer and bonds redistribution in formed transition complex leads to break of cycle configuration with formation of OAc^- ion, acetaldehyde, elimination of CO and is completed by formation of catalytic particles with mixed ligands of general formula $Ni_x(acac)_y(L^1{}_{ox})_z(L^2)_n$ ($L^1{}_{ox}$= $MeCOO^-$) ("A") [29,30]. Transformation of complexes $Ni(acac)_2L^2$ (L^2=MP) leads to formation of binuclear complexes "A" with formula: $Ni_2(OAc)_3(acac)L^2$ (Scheme 1) [29]. The structure of the last ones is proved kinetically and by various physical-chemical methods of analysis (mass-spectrometry, electron and IR-spectroscopy, elemental analysis).

$$L^2{\cdot}L^1Ni(COMeCHMeCO)_2+O_2 \rightarrow L^2{\cdot}L^1Ni(COMeCHMeCO)^+\ldots O_2^-$$
$$L^2{\cdot}L^1Ni(COMeCHMeCO)^+\ldots O_2^- \rightarrow L^2{\cdot}L^1Ni(MeCOO)+MeCHO+CO$$
$$L^1=(COMeCHMeCO)^- 2\ Ni(COMeCHMeCO)_2\ L^2 \xrightarrow{O_2}$$
$$Ni_2(MeCOO)_3(COMeCHMeCO)L^2+3MeCHO+3CO+L^2$$

L^2=N-methylpyrrolidone-2

Scheme 1.

Transformation of $Ni(L^1)_2$ (L^1=enamac$^-$, chelate group (O/NH)) is realized in the absence of activating ligands (L^2) [30] ($L^1{}_{ox}$=NHCOMe$^-$ or MeCOO$^-$) (Scheme 2).

$$Ni(COMeCHMeCNH)_2 + O_2 \longrightarrow L{\cdot}Ni(COMeCHMeCNH)^+\ldots O_2^-$$
$$L{\cdot}Ni(COMeCHMeCNH)^+\ldots O^-_2 \longrightarrow L{\cdot}Ni(NHCOMe) + MeCHO + CO$$

(Q)

$\downarrow H_2O$

$$L{\cdot}Ni(MeCOO) + NH_3$$

(P)

where L=(COMeCHMeCNH)$^-$.

Scheme 2.

Similar change if complexes' ligand environment under the action of O_2 was observed in reactions of oxygenation imitating the action of dioxygenases (quercetenase, L-tryptophan-2,3-dioxygenase) [31, 32].

Transformation of $Fe(II)(acac)_2 \cdot L^2$ complexes resulted from catalyzing $\{Fe(III)(acac)_3+L^2\}$ ethylbenezene oxidation (80°C, 120°C) proceeds by analogous mechanism. Probable structure of transformation products is as follows:

$Fe(II)_x(acac)_y(Oac)_z(L^2)_n(L^2=DMFA))$ [25, unpublished data].

Mechanism of cobalt (II) complexes transformation in the course of ethylbenzene oxidation (120°C) is different. On the base of obtained in [29] experimental data they suggested the transformation mechanism by analogy with known from literature mechanism as a result of interaction of $Co(II)(acac)_2 \cdot L^2$ with peroxide radicals with formation of catalytically active complexes of the following probable structure $[Co(III)(L^1)_2 \cdot L^2 \cdot (RO_2^-)]$ [33] (see below).

The established mechanism of $\{Ni(L^1)_2+L^2\}$ transformation into more active selective catalysts allowed solution of the problem of selective oxidation. Methods of ethylbenzene oxidation optimization in the presence of $\{Ni(L^1)_2+L^2\}$ catalytic systems were proposed and realized. Some of them are connected with increase of concentration of active mixed ligand complexes $Ni_x(acac)_y(L^1{}_{ox})_z(L^2)_n$ ("A") or stability of catalyst's active form, with inhibition of process of complete oxidation of "A" into inactive homogeneous-ligand complexes $Ni(MeCOO)_2$ or $Ni(NHCOMe)_2$ carrying out heterolysis of PEH [29, 34, 35].

1. Introduction of Activating Additives in the Course of Catalyzed Oxidation Process

It is evident from analysis of scheme of catalyzed oxidation including participation of catalyst in chain initiation reaction under catalyst interaction with ROOH and also in chain propagation ($Cat + RO_2^{\cdot} \rightarrow$) that with decrease of $[Cat]^0$ the rate of reaction should be decreased, and $[ROOH]^{max}$ should be increased [2]. In this connection we may expect that decrease of $[M(acac)_n]^0$ will lead to increase of ethylbenzene oxidation selectivity into PEH also in the absence of L^2.

In work [25] they established that at low concentrations $[Ni(acac)_2]=(0,5\div1,5)\cdot10^{-4}$M (120°C) in the absence of L^2 high values of selectivity of catalyzed oxidation are possible: $S_{PEH}^{max} = 90\%$, but only at insignificant conversion degree, C = 4-6%. Products AP and MPC (P) are formed in this case not from PEH but parallel with PEH, i.e. $w_P / w_{PEH} \neq 0$ at $t \rightarrow 0$, and furthermore $w_{AP} / w_{MPC} \neq 0$ at $t \rightarrow 0$ that indicates on parallelism of formation of AP and MPC (P = AP or MPC) [30, 35]. At these conditions addition of electron-donor monodentate ligands turned to be low effective [25] and the change of S_{PEH}^{max} and $C_{S=90\%}$ under introduction of additives L^2 (L^2 = HMFA) into system practically was not observed.

For increase of conversion degree of oxidation at maintaining of S_{PEH}^{max} not less than 90% the method of catalytic system activation in developed oxidation process was used [29]. Application of the method of influence on chemical reaction not only at its beginning but also on various stages of process is one of the methods of optimization of complex multi-stage oxidation reaction [36, 37].

In the case of introduction of $Ni(II)(acac)_2$ into ethylbenzene oxidation reaction catalyzed by {$Ni(II)(acac)_2(1,5\cdot10^{-4}$ M)+HMPA($1,0\cdot10^{-3}$ M)} catalytic system the maximum possible conversion degree C at which reaction selectivity is not less than S_{PEH} = 90% is significantly increased (in ~ 3 times) (Figure 1) [29].

The necessary condition is introduction of additives at conditions of reaction stationarity by S_{PEH} ($S_{PEH} = S_{PEH}^{max}$). Reactivation of catalytic system is not observed when one introduces additional amount of catalyst on sections corresponding to decrease of oxidation selectivity ($S_{PEH} < S_{PEH}^{max}$ at C < 4% and C > 5-6%).

Mechanism of activation consists in increase of stationary concentration of catalytically active complex "A" responsible for selectivity of oxidation process. Activation proceeds in two stages:

1. Interaction of $Ni(acac)_2{\cdot}L^2$ with $Ni(OAc)_2$ with formation of complex with mixed ligands $Ni(acac)(OAc){\cdot}L^2$:

$$Ni(II)(acac)_2{\cdot}L^2 + Ni(II)(OAc)_2 \longrightarrow Ni(II)(acac)(OAc){\cdot}L^2 \quad (1)$$

2. Further oxidation of this complex to binuclear one $Ni_2(acac)(OAc)_3{\cdot}L^2$ ("A"). Higher activity of multi-ligand nickel complexes in reactions of connection of electrophyls E by γ-C-atom of β-diketonate ligand is reported in [38]:

$$Ni(II)(acac)(OAc){\cdot}L^2 + O_2 \longrightarrow Ni_2(acac)(OAc)_3{\cdot}L^2 \quad (2)$$

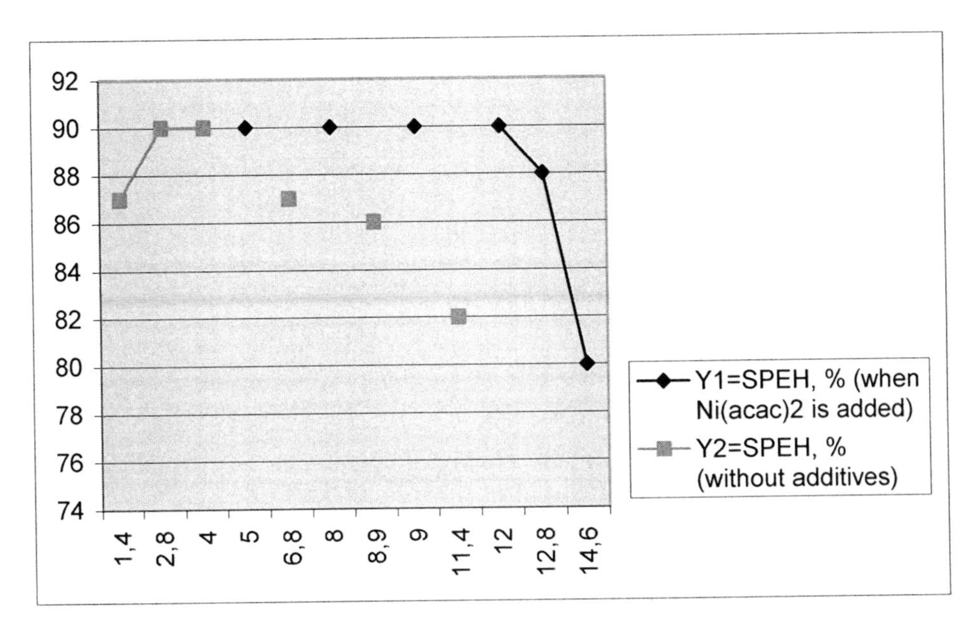

Figure 1. dependences of selectivity of ethylbenzene oxidation into PEH (SPEH) conversion level of ethylbenzene (CPEH) in the presence of system {Ni(II) добавок 4,5·10-4 M Ni(II)(acac)2 [Ni(II)(acac)2]0=1,5·10-4 M, [HMPA]=1,0·10-3 M, 120°C.

Activation of catalytic system $\{Ni(II)(acac)_2+L^2\}$ by acacH additives testifies to favour of exchange interaction between $Ni(II)(acac)_2 \cdot L^2$ and $Ni(II)(OAc)_2$ (1). However, conversion degree $\mathbf{C}_{S=90\%}$ is increased in this case to a lower extent than under addition of $Ni(acac)_2$ into system, obviously due to fast oxidation of non-coordinated acetylacetonate [39].

Role of reaction (2) in catalytic system activation mechanism is proved under application of method of electron-acceptor additives.

We established [29] that effect of conversion degree increase $C_{S=90\%}$ is not observed under combined introduction of $Ni(acac)_2$ and π-acceptors (tetrtacyanethylene (TCE) or chloranil (CA)) into system. Since electron affinity is increased in the raw O_2(0,87[40]) <CA(1,3[41]) < TCE(1,6[42], 2,2[41]) the observed passivation of catalytic system by additives of CA and TCE is obviously connected with the fact that stronger in comparison with O_2 electron acceptors TCE and CA electrophylicaly attack acetylacetonate ligand by γ-C-atom forming out-spherical complexes and hinder connection of O_2 by this bond and consequently the oxidation of $(acac)^-$ to acetate-ion.

2. Application of Additives of Macro-Cycle Ethers and Ammonium Quaternary Salts as Ligands-Modifiers for Increase of $M(Acac)_N$ Activity in Reaction of Selective Oxidation of Ethylbenzene to PEH

Interest in studying of structure and catalytic activity of nickel complexes (especially nickel complexes with macrocycle ligands) is increased recently in connection with discovering of nickel-containing ferments [43, 44]. So, they established that active sites of ferment urease are binuclear nickel complexes containing N/O-donor ligands. Cofactor of oxidation-reduction ferment methyl-S-coenzyme-M-reductase in structure of methanogene bacteria is tetra-aza-macrocycle nickel complex with hydrocorfine $Ni(I)F_{430}$ axially coordinated inside of ferment cavity [44].

Inclusion of transition metals cations into cavity of macrocycle polyethers is proved by now by various physical-chemical methods. Concrete structure of complex at that is determined not only by geometric accordance of metal ion and kraun-ether cavity but by the whole totality of electron and spatial factors created by metal, polyether and other ligands atoms and also by solvent [45].

On the base of literature data we may expect the increase of conversion degree C of catalyzed by $Ni(L^1)_2$ ($1,5 \cdot 10^{-4}$ M) ethylbenzene oxidation (120°C) at conservation of S_{PEH}^{max} not less than 90% in the presence of additives 18K6. Coordination of 18K6 with $Ni(II)(acac)_2$ may promote oxidative transformation of nickel (II) complexes (schemes 13.1 and 13.2) into catalytically active particles. For example, the ability of kraun-ethers to catalyze electrophilic reactions of connection to γ-C-atom of $acac^-$-ligand is known [46, 47]. Steric factors appearing under coordination of 18K6 may hinder transformation of catalytically active multi-ligand complex into inactive particles.

The ability of ammonium quaternary salts to complexation with transition metals compounds is also known. They proved for example that $M(acac)_2$ (M=Ni, Cu) form with R_4NX ($X=(acac)^-$, R=Me) complexes of $[R_4N][M(acac)_3]$ structure. Spectral proofs of octahedral geometry for these complexes were got [48]. Complexes Me_4NiBr_3 were synthesized and their physical properties were studied [49].

Furthermore, it is known that R_4NX in hydrocarbon mediums forms with acetylacetone complexes with strong hydrogen bond $R_4N^+(X...HOCMe{=}CHCOMe)^-$ in which acetylacetone is totally enolyzed [50].

We may expect the increase of $M(acac)_n$ efficiency in ethylbenzene oxidation into hydroperoxide in the presence of R_4NX by the mechanism presented in Scheme 1. The controlled by R_4NX regio-selective connection of O_2 by γ-C-atom of $(acac)^-$ ligand in complex $M(acac)_n{\cdot}R_4NX$ is probable enough. Various electrophilic reactions in complexes $R_4N+(X...HOCMe{=}CHCOMe)^-$ proceed by γ-C-atom of acetylaceton [47, 50].

However, introduction of 18K6 or Me_4NBr additives into ethylbenzene oxidation reaction catalyzed by complexes $Ni(L^1)_2$ at conditions mentioned above lead to unexpected results. Significant increase of conversion degree of oxidation into PEH at maintenance of selectivity on level S_{PEH}=90-80%, increase of S_{PEH}^{max} and initial rate of reaction w^0 were observed. In the presence of 18K6 the highest values of w^0 (Figure 2) and also of S_{PEH} and C are reached at ratios [Cat]:[18K6]=1:1 ($Ni(O/NH)_2$) and [Cat]:[18K6]=1:1 and 1:2 ($Ni(O/O)_2$). At that degree of conversion into PEH at maintenance of selectivity S_{PEH} at not less than 90% is increased from 4-6 up to 12% for ($Ni(O/O)_2$) and from 12 up to 16% for ($Ni(O/NH)_2$). As it is obvious the effect of influence of 18K6 on S_{PEH} and C is higher in the case of complex with chelate group (O/O). At that S_{PEH}^{max} =98% is observed not in developed reaction of oxidation as under catalysis of $Ni(O/NH)_2$ (S_{PEH}^{max}=94%) but at the beginning of process. In the case of additives of Me_4NBr into reaction of ethylbenzene oxidation catalyzed by $Ni(II)(acac)_2$ the value of S_{PEH}^{max}=95% is higher than under catalysis by $Ni(II)(acac)_2$ with out addition of L^2. S_{PEH}^{max} is reached not at the beginning of reaction of ethylbenzene oxidation but at C=2-3%. Selectivity remains in the limits $90\% < S \leq 95\%$ to deeper transformation degrees of ethylbenzene C≈19% than in the presence of additives 18K6 (C≈12%).

Additives of 18K6 or Me_4NBr to ethylbenzene oxidation reaction catalyzed by $Ni(L^1)_2$ lead to significant hindering of heterolysis of PEH with formation of phenol (P.) responsible for selectivity decrease. At that induction period of P. formation in the presence of Me_4NBr is significantly higher than in the case of 18K6.

Influence of quaternary ammonium salt on catalytic activity of $Ni(II)(acac)_2$ as selective catalyst of ethylbenzene oxidation into PEH extremely depends on radical R structure of ammonium cation. If cetyltrimethylammonium bromide (CTAB) is added S_{PEH}^{max} is reduced down to 80-82% [34]. w^0 is significantly increased, in ~ 4 times in comparison with ethylbenzene catalysis by $Ni(II)(acac)_2$ complex. Initial rate of PEH accumulation w_{PEH}^0 is higher than in the case of ehtylbenzene oxidation catalyzed by the system {$Ni(II)(acac)_2$ + Me_4NBr}. However initial rates of accumulation of side products of reaction of AP with MPC are also significantly increased. Selectivity decrease connected with heterolysis of PEH with phenol formation is observed at lower conversions of RH transformation.

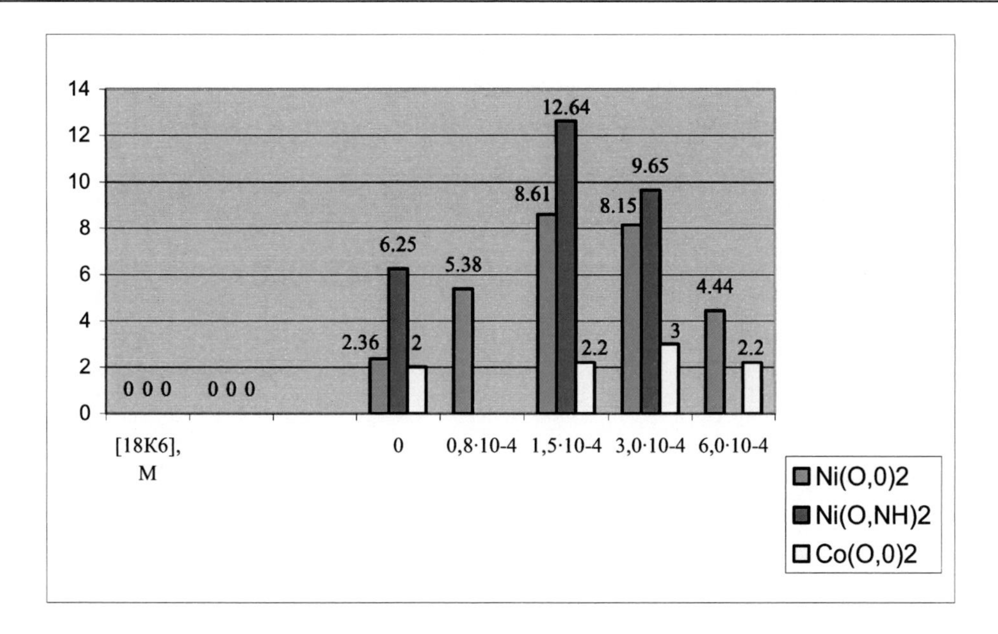

Figure 2. Dependences of initial rates w^0 ($M \cdot sec^{-1}$) of ethylbenzene oxidation reactions catalyzed by $\{M(L^1)_2+18K6\}$ (M=Ni(II), Co(II), L^1=$acac^{-1}$, $enamac^{-1}$) on concentration of [18K6] (M). $w^0 \cdot 10^4$ (M=Co(II)), $w^0 \cdot 10^5$ (M=Ni(II)), $[M(L^1)_2]=1{,}5 \cdot 10^{-4}$ M, 120°C.

Analysis of consequence of ethylbenzene oxidation products formation catalyzed by systems $\{Ni(II)(acac)_2+18K6\}$ and $\{Ni(II)(acac)_2+R_4NBr\}$ showed that in the course of reaction of ethylbenzene oxidation PEH, AP and MPC were formed parallel ($w_P/w_{PEH} \neq 0$ at $t \to 0$), AP and MPC were formed also parallel ($w_{AP}/w_{MPC} \neq 0$ at $t \to 0$).

Catalysis of ehtylbenzene oxidation initiated by $\{Ni(II)(acac)_2$ + CTAB$\}$ system is not connected with formation of micro-phase by the type of inverse micelles since the micellar effect of CTAB revealing at $t^0 < 100^0$ [52] is as a rule not important at $t^0 \geq 120^0$. Furthermore, as we saw the system $\{Ni(II)(acac)_2$ + CTAB$\}$ was not active in decomposition of ROOH.

The observing significant effects of S_{PEH}, C and w^0 increase are not connected with influence of kraun-ether or Me_4NBr. We established that in the presence of one 18K6 or R_4NBr without nickel complex auto-catalytic developing of process with initial rates by order lower was observed, and selectivity of process by PEH equal at the beginning of reaction to 85% (18K6) or 95% (Me_4NBr) was sharply reduced with the increase of ethylbenzene conversion degree. Formation of P. is observed at that from the beginning of reaction.

For estimation of catalytic activity of nickel (and also cobalt) complexes as selective catalysts of ehtylbenzene oxidation into α-phenylethylhydroperoxide we proposed to use parameter $\tilde{S} \cdot C$. $\tilde{S}$ is averaged selectivity of oxidation into PEH characterizing change of S in the course of oxidation from S^0 at the beginning of reaction to some S^{lim} conditional value chosen as a standard. For comparable by value systems selectivity as S^{lim} was selected the value S^{lim} = 80% approximately equal to selectivity of non-catalyzed ethylbenzene oxidation into PEH at initial stages of reaction. At that the cases when $S^0 > S^{lim}$ (nickel (II) complexes) and $S^0 \leq S^{lim}$ (nickel (II) and cobalt (II) complexes) are possible; C – conversion degree at $S = S^{lim}$ [30, 51].

As it is obvious (Table 1) the system {$Ni(acac)_2$+Me_4NBr} by the value of parameter $\tilde{S}\cdot C$ is the most active catalyst of ehtylbenzene oxidation into PEH. System {$Ni(L^1)_2$+18K6} is more selective catalyst in comparison with $Ni(L^1)_2$, and systems {$Ni(acac)_2$+18K6} and {$Ni(enamac)_2$+18K6} with ratio [Cat]:[18K6]=1:1 are slightly differed in the value of parameter S·C. In the absence of L^2 complex $Ni(enamac)_2$ is more selective catalyst in comparison with $Ni(acac)_2$.

Table 1. The value $\tilde{S}\cdot C$ in reactions of ethylbenzene oxidation catalyzed by nickel and cobalt complexes. [Cat] = $1,5\cdot10^{-4}$ M, 120°C.

Cat	$\tilde{S}\cdot C\cdot10^{-2}$(%,%)
$Ni(enamac)_2$	15,9
{$Ni(enamac)_2$+18K6} (1:1)	21,2
$Ni(acac)_2$	9,6
{$Ni(acac)_2$+18K6} (1:1)	20,6
{$Ni(II)(acac)_2$+Me_4NBr} (1:1)	24,3
$Ni(NO_3)_2\cdot18K6\cdot6H_2O$(1)	16,0
$Co(acac)_2\cdot18K6_2$	9,9
$2Co(NO_3)_2\cdot18K6\cdot6H_2O$(2)	15,8

Observing synergetic effects of w^0 (Figure 2, Table 2) and parameter $\tilde{S}\cdot C$ (Table 1, Figure 3) increase testify to formation of active in oxidation complexes $Ni(L^1)_2$ with (L^2) [53] of structure 1 : 1 and 1 : 2 (L^2=18K6) or structure 1 : 1 (L^2=Me_4NBr) and also the products of their transformation [34, 55].

Table 2. Rates of accumulation of ethylbenzene oxidation products at the beginning of reaction (w^0) and in the course of process (w) ($M\cdot c^{-1}$), and calculated rates of initiation (w^0_i and w_i) and chain propagation (w^0_{pr} and w_{pr}). Catalysis by complexes $Ni(L^1)_2$ (L^1 = $enamac^-$, $acac^-$) and $Ni(L^1)_2\cdot L^2_n$ (L^2 = 18K6, n = 1,2). [$Ni(L^1)_2$] = $1.5\cdot10^{-4}$ M, 120^0 C.

$Ni(L^1)_2$	$[L^2]\cdot10^4$	$w^0_{PEH}\cdot10^5$	$w^0_{AP+MPC}\cdot10^6$	$w_{PEH}\cdot10^5$	$w_{AP+MPC}\cdot10^6$	$w^0_i\cdot10^6$	$w^0_{pr}\cdot10^6$	$w_i\cdot10^7$	$w_{pr}\cdot10^7$
$Ni(O,NH)_2$	—	5,5	7,5	1,1	0,9	2,4	5,1	1,1	8,4
	1,5(1:1)	11,7	9,44	1,0	1,4	10,5	—	0,8	14,2
	3,0(1:2)	8,88	7,72	1,0		6,04	1,18	0,8	15,2
Without catalyst	—	—	—	1,0	6,3	—	—	—	—
$Ni(O,O)_2$	—	2,1	2,6	1,1	1,1	0,3	2,3	—	—
	1,5(1:1)	8,12	4,9	1,6	1,3	5,1	—	—	—
	3,0(1:2)	7,7	4,5	1,4	2,0	4,5	—	—	—

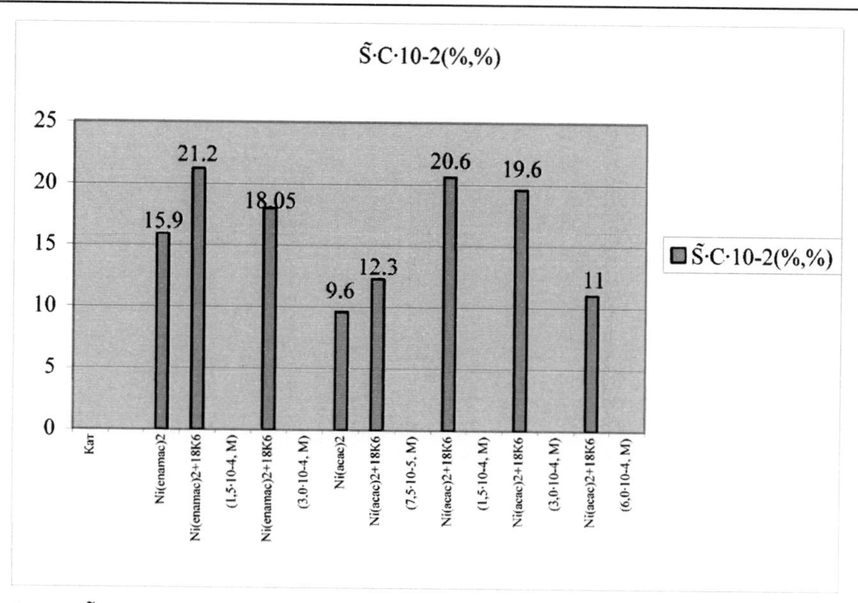

Figure 3. Value S̃·C in reactions of ehtylbenzene oxidation catalyzed by $Ni(acac)_2$ and $Ni(enamac)_2$ complexes without additives and in the presence of L^2 (L^2=18K6, M) [Cat]=$1{,}5\cdot10^{-4}$ M, 120°C.

We established that in the case of catalysis {$Co(acac)_2$+18K6} the optimal by parameter S̃·C catalytic system {$Co(acac)_2(1{,}5\cdot10^{-4}$ M)+18K6($3\cdot10^{-4}$ M)} was less active catalyst of ethylbenzene oxidation into PEH in comparison with {$Ni(L^1)_2$+18K6} (Table 1). Although additives 18K6 cause significant increase of S_{PEH}^{max} from 56% ($Co(acac)_2$) up to 80% ({$Co(acac)_2$+18K6}). w^0 is increased in ~1,5 times at [18K6]:[$Co(acac)_2$]=2:1 ($(w^0)^{max}$) (Figure 13.2) [33].

Formation of complexes between $M(L^1)_2$ (M=Ni(II), Co(II)) and 18K6, $Ni(acac)_2$ with R_4NBr was also proved by spectrophotometry under analysis of UV spectra of adsorption of $Ni(L^1)_2$ and 18K6 (R_4NBr), $Co(acac)_2$ and 18K6mixtures solutions. At that 18K6 or R_4NBr coordinate with metal ion with preservation of ligand L^1 in internal coordination sphere of complex [33-35]. Under formation of complexes of $Ni(L^1)_2$ with R_4NBr in spite of axial coordination of anion Br^- by the fifth coordination place of nickel (II) ion the out-spherical coordination of ammonium cation with acetylacetonate-ion is also possible [34].

Fe(III)$(acac)_3$ and formed in the course of oxidation Fe(II)$(acac)_2$ are inactive in PEH destruction [54]. In ethylbenzene reaction oxidation catalyzed by Fe(III)$(acac)_3$ ([Cat]=$(0{,}5\div7{,}5)\cdot10^3$ (M·), 80 and 120°C [54]) the main products of oxidation PEH, MPC and AP are formed parallel as at the beginning of reaction, so at deeper stages of oxidation [54]. In the presence of additives of electron-donor monodentate ligands (HMPA, DMPA) S_{PEH}^{max} is increased from 42 up to 55-60% ([Cat] = $5\cdot10^{-3}$ M) [25, unpublished data].

In ethylbenzene oxidation reaction in the presence of {Fe(III)$(acac)_3(5\cdot10^{-3}$ M)+18K6} (80°C) the S_{PEH}^{max} reached in developed process is higher than in the case of additives of monodentate ligands: S_{PEH}^{max} = 70% ($[18K6]^0 = 0{,}5\cdot10^{-3}$ M) and S_{PEH}^{max} = 75,7% ($[18K6]^0 = 5\cdot10^{-3}$ M). Conversion degree in increased in 2 times from C = 4 up to 8% (at S_{PEH} =40-(70-75,7)%). selectivity growth occurs at the expense of significant decrease of AP and MPC formation rate at parallel stages of propagation and quadratic termination of chain [55].

In the presence of {Fe(III)$(acac)_3$+18K6} synergetic effect of S̃·C parameter increase is observed, the growth in 2,5 and 2,8 times at $[18K6]^0 = 0{,}5\cdot10^{-3}$ M and $18K6]^0 = 5\cdot10^{-3}$ M correspondingly in comparison with catalysis by Fe(III)(acac)$_3$ ($\tilde{S}\cdot C=2{,}1\cdot 10^2$ (%,%)) [55]. In given case for value S^{lim} they accept S^{lim} =40%, approximately equal to S_{PEH} for Fe(III)(acac)$_3$ in developed reaction, C – conversion degree at which $S_{PEH} \leq 40\%$. Obtained kinetic regularities of ethylbenzene oxidation testify to formation presumably of $(Fe(II)(acac)_2)_x\cdot(18K6)_y$ [45] and products of their transformation in the course of complexes oxidation.

Analogous effects of growth of $\tilde{S}\cdot C$ (in ~2,2–2,4 times) were observed under the use as L^2 additives of $(C_2H_5)_4NBr$ and other R_4NBr [unpublished data]. S_{max} = 60-63% in the case of catalysis by systems {Fe(III)(acac)$_3$ + R_4NBr} is lower than maximum S_{PEH} = 75,7% reached under catalysis by {Fe(III)(acac)$_3$ + 18K6 ($5\cdot 10^{-3}$ M)} in developed process [55].

2.1. Participation of Catalysts Active Forms in Elementary Stages of Radical-Chain Ethylbenzene Oxidation Catalyzed by {M(L^1)$_2$+18K6} (M=Ni(II), Fe(III), L^1=acac^{-1} and by Complexes of Ni (II) and Co (II) Salts with 18K6

We propose the method for estimation of catalytic activity of complexes formed *in situ* at the beginning of reaction and in developed process, at elementary stages of oxidation process [30, 34, 35, 51] by simplified scheme assuming quadratic termination of chain and equality to zero of rate of homolytic decomposition of ROOH. In the framework of radical-chain mechanism the chain termination rate in this case will be:

$$w_{term}=k_6[RO_2^{\cdot}]^2=k_6\left\{\frac{w_{PEH}}{k_2[RH]}\right\}^2 \quad (1)$$

where w_{PEH} – rate of PEH accumulation, k_6 – constant of reaction rate of quadratic chain termination; k_2 – constant of rate of chain propagation reaction $RO_2^{\cdot}$ + RH→.

We established that complexes $M(L^I)_n$ (M=Ni(II), Fe(III)) were inactive in PEH homolysis, products MPC and AP were formed at stages of chain propagation Cat + $RO_2^{\cdot}$→ and quadratic termination of chain. Actually, w^0~$[Cat]^{1/2}$ and w_i^0~[Cat] and linear radicals termination on catalyst may be not taken into account. In the case of quasi-stationarity by radicals $RO_2^{\cdot}$ the values $w_{term.}=w_i$ are the measures of nickel complexes activity in relation to molecular O_2. Discrepancy between w_{AP+MPC} and w_{term} (Tables 2 and 3) in the case of absence of linear termination of chain is connected with additional formation of alcohol and ketone at the stage of chain propagation Cat + $RO_2^{\cdot}$→:

$$w_{pr.}= w_{AP+MPC} - w_{term} \quad (2)$$

The direct proportional dependence of w_{pr}^0 on [Cat] testifies in favour of nickel and iron complexes participation at stage of chain propagation Cat + $RO_2^{\cdot}$→.

We suggest that these conditions w^0~$[Cat]^{1/2}$ and w_i^0~[Cat] will be fulfilled also in the presence of additives of 18K6 or R_4NBr.

Except theoretical consideration in [18], activity of transition metals complexes $M(L^1)_n$ ($M=Ni$, Co, Fe, L^1 = enamac$^-$, acac$^-$) at stage of chain propagation ($Cat + RO_2^{\cdot} \rightarrow$) of ethylbenzene oxidation is estimated only in our works [30, 34, 35, 51, 55].

Table 3. The values of $\tilde{S} \cdot C$, $C_{S=80\%}$, $[PEH]^{max}$ in reactions of ethylbenzene oxidation in the presence of triple catalytic systems {$Ni(II)(acac)_2$+L^2+P.}. $[Ni(II)(acac)_2]=3{,}0 \cdot 10^{-3}$ M, 120°C.

$[L^2]$, M	[P.], M	$\tilde{S} \cdot C \cdot 10^{-2}$,(%,%)	C, %	$[PEH]^{max}$(mass%)
$3{,}0 \cdot 10^{-3}$ (NaSt)	$3{,}0 \cdot 10^{-3}$	30,1	>35,0	27
$3{,}0 \cdot 10^{-3}$ (NaSt)	$1{,}5 \cdot 10^{-3}$	20,0	25,0	21
$3{,}0 \cdot 10^{-3}$ (NaSt)	—	10,9	14,2	16
$3{,}5 \cdot 10^{-3}$ (LiSt)	$1{,}0 \cdot 10^{-3}$	20,6	24,0	26,0
$3{,}5 \cdot 10^{-3}$ (LiSt)	$2{,}0 \cdot 10^{-3}$		22,0	22,0
$3{,}5 \cdot 10^{-3}$ (LiSt)	—	14,4	18,0	19,0
$2{,}0 \cdot 10^{-2}$ (HMPA)	$1{,}0 \cdot 10^{-3}$	16,2	19,0	20,0
$2{,}0 \cdot 10^{-2}$ (HMPA)	$5{,}0 \cdot 10^{-4}$	13,6	16,0	20,0
$2{,}0 \cdot 10^{-2}$ (HMPA)	—	11,1	14,0	17,5
$7{,}0 \cdot 10^{-2}$ (MP)	$3{,}0 \cdot 10^{-3}$	17,5	21,0	21,0
$7{,}0 \cdot 10^{-2}$ (MP)	$4{,}6 \cdot 10^{-4}$	18,1	21,0	21,0
$7{,}0 \cdot 10^{-2}$ (MP)	—	11,9	12,2	17,5

Investigation of reaction ability of peroxide complexes [LM-OOR] (M=Co, Fe) preliminary synthesized by reactions of compounds of Co and Fe with ROOH or $RO_2^{\cdot}$-radicals [56-58] confirms their participation as intermediates in reactions of hydrocarbons oxidation. Obviously, the schemes of radical-chain oxidation including intermediate formation [LM-OOR] [56-59] with further homolytic decomposition of peroxo-complexes ($[LM\text{-}OOR] \rightarrow R'C{=}O\ (ROH) + R^{\cdot}$) may explain parallel formation of alcohol and ketone under ethylbenzene oxidation in the presence of $M(L^1)_n$ (L^1 = enamac$^-$, acac$^-$) and their complexes with 18K6 and R_4NBr.

We established that mechanism of selective catalysis depended on both activity of Cat in PEH decomposition and also on ratio of rates of chain initiation w_i (activation by O_2) and propagation (w_{pr}).

In works [14, 18] N.M. Emanuel and D. Gal show that when in ethylbenzene oxidation reaction at high temperatures the formation of active sites occurs in reaction of chain initiation ($RH+O_2 \rightarrow$) and under chain decomposition of PEH, the value S_{PEH} to a significant extent should be determined by factor of instability of PEH $\beta = w_{PEH}^- / w_{PEH}^+$ (w_{PEH}^- – sum rate of PEH decomposition (thermal (molecular) and chain), w_{PEH}^+ – rate of chain formation). Actually, it turned out that value β in the course of non-catalyzed process of ethylbenzene oxidation is increased at the expense of rise of PEH chain decomposition rate that leads to reduction of S_{PEH} [14, 18].

At conditions of ethylbenzene oxidation catalyzed by $Ni(L^1)_2$, $Ni(L^1)_2 \cdot 18K6_n$, $Ni(acac)_2 \cdot R_4NBr$ the value is $\beta = w_{PEH}^- / w_{PEH}^+ \rightarrow 0$ as at the beginning, so in developed process, the direction of AP and MPC formation is changed (consequent → parallel), S_{PEH}

depends on catalyst activity at stages of chain initiation (activation by O_2) and propagation Cat+ $RO_2^{\cdot}$ →.

Calculation by formulas (1) and (2) show that high activity of "primary" complexes $Ni(II)(L^I)_2 \cdot 18C6_n$ as selective catalysts of ethylbenzene into PEH oxidation is connected with five-(chelate group (O/NH)) and twenty-(chelate group (O/O))-fold growth of rate w_i^0 in comparison with catalysis by $Ni(L^I)_2$, hindering of rate of chain propagation (w_{pr}^0) Cat+$RO_2^{\cdot}$ → (Table 2), and also homolysis and heterolysis of PEH. Under catalysis by complexes $Ni(acac)_2 \cdot 18C6_n$ (n=1,2) and $Ni(enamac)_2 \cdot 18C6_n$ (n=1) experimentally determined w_{AP+MPC}^0 completely concide with calculated ones by formula (1) values of w_{term}^0 (Table 2).

Complexes $Ni(O,NH)_2 \cdot 18C6_n$ are twice as more active than $Ni(O,O)_2 \cdot 18C6_n$ at stage of free radicals origin (w_i^0), although the "kraun-effect" (increase of w^0 and w_i^0 under the effect of 18K6 additives) observed in the case of catalysis by $Ni(II)(O,NH)_2 \cdot 18K6_n$ is lower. It may be explained by reduction of acceptor properties of complex $Ni(II)(O,NH)_2$ in comparison with $Ni(II)(O,O)_2$ in relation to coordination 18K6 that is caused by covalent character of bonds Ni-NH and reduction of effective charge of metal ion.

In developed oxidation reaction catalyzed by $Ni(L^I)_2 \cdot 18C6_n$ (and also by $Ni(acac)_2 \cdot R_4NBr$) when the reduction of w in comparison with w^0 is observed, at slightly changed S_{PEH} constancy of w is remained right up to the moment of phenol formation coinciding with sharper decrease of S_{PEH}. Such regularities of S and w change in the course of oxidation testify in favour of initial complexes transformation into novel catalytic particles, probably in accordance with schemes 1 and 2.

Conditions allowing estimation of w_{pr} and w_i (formulas 1 and 2) in developed process under catalysis $Ni_x(L^I)_y(L^I_{ox})_z$ и $Ni_x(L^I)_y(L^I_{ox})_z \cdot 18K6_n$ (L^I= enamac^{-1}) are fulfilled. As it is obvious from Table 2 the role of reaction of chain propagation in developed oxidation reaction of ethylbenzene is increased.

In contrast to catalysis by complexes $Ni(II)(L^1)_2$ (L^1=acac^{-1}, enamac^{-1}) with 18K6 in reaction of ethylbenzene oxidation catalyzed by $Ni(II)(L^1)_2$ in the absence of kraun-ethers additives increase of initial rate of oxidation is connected mainly with participation of catalyst at stage of chain propagation. At that under catalysis by $Ni(O,NH)_2$ complex the value w^0_{pr} is twice as much than under catalysis by $Ni(O,O)_2$. at the same time the rate of chain initiation almost in order exceeds w^0_i in oxidation reaction catalyzed by $Ni(II)(acac)_2$. As it obvious, presence of donor NH-groups in chelate group of nickel complex promotes significant increase of role of activation reaction of molecular oxygen in catalysis mechanism. Reduction of S_{PEH} coinciding with significant rise of P. accumulation rate may be explained by complete oxidation of active multi-ligand complexes to mono-ligand $Ni(MeCOO)_2$ or $Ni(NHCOMe)_2$ carrying out PEH heterolysis [29, 30, 35].

Under catalysis of ethylbenzene oxidation by system {$Ni(II)(acac)_2$+Me_4NBr} the complexes $Ni(II)(acac)_2 \cdot Me_4NBr$ formed in situ at initial stages of reaction reveal less activity at stage of free radicals origin (activation by O_2) and higher activity at stage of chain propagation Cat+ $RO_2^{\cdot}$ → in comparison with complexes $Ni(II)(L^1)_2 \cdot 18K6_n$ (Figure 4). Ratio $w^0_{pr}/w^0_i \approx 2$ coincides with analogous parameter for $Ni(II)(enamac)_2$. Obviously, due to this reason the value of S_{PEH}^0 under catalysis by $Ni(II)(acac)_2 \cdot Me_4NBr$ inactive in PEH homolysis is lower than S_{PEH}^0 in the case of $Ni(II)(L^1)_2 \cdot 18K6_n$.

Under substitution of radical CH_3 in cation R_4N^+ by radical n-$C_{16}H_{33}$ the value of S^0_{PEH} is reduced. Activity of formed complexes $Ni(II)(acac)_2 \cdot CTAB$ at stages of chain initiation and propagation is increased in 4,6 and in 20,5 times correspondingly. At that the rate of PEH accumulation (w^0_{PEH}) is increased only in 2 times, and w^0_{AP+MPC} in 15,4 times in comparison with catalysis by $Ni(II)(acac)_2 \cdot Me_4NBr$.

Higher value of $\tilde{S} \cdot C$ parameter in the case of ethylbenzene oxidation catalysis by systems $\{Ni(II)(acac)_2 + Me_4NBr\}$ in comparison with $\{Ni(II)(L^1)_2 + 18K6\}$ (Table 1) may be explained by higher stability of formed in the course of reaction catalytic particles, presumably $Ni_x(acac)_y(OAc)_z \cdot Me_4NBr$ to complete oxidation to nickel acetate. Out-spherical coordination of Me_4NBr creating sterical hindrances for regio-selective oxidation of $(acac)^-$–ligand may be favourable to the last fact [34].

Maintenance of high selectivity of developed reaction catalyzed by systems $\{Ni(II)(L^1)_2 + 18K6\}$ and $\{Ni(II)(L^1)_2 + R_4NBr\}$ in comparison with non catalyzed oxidation of ethylbenzene ($S_{PEH} << 80\%$) is connected mainly with hindering of chain and heterolytic decomposition of PEH and change of products formation mechanism (consequent→parallel).

Complexes $Ni(NO_3)_2 \cdot 18K6 \cdot 6H_2O$ (1) are less selective catalysts in comparison with acetylacetonate and enaminoacetonate complexes $Ni(L^I)_2 \cdot 18C6_n$. In the course of oxidation the transformation of complexes (1) was not observed [51]. Obviously, the presence of $(acac)^-$ or $(enamac)^-$– ligands is necessary for high catalytic activity of nickel complexes with 18K6. In this case the conditions allowing estimation of w_i^0 and w_{pr}^0 by formulas (1) and (2) are also fulfilled. It turned out that complexes (1) are less active at initial stages of oxidation in reaction of oxygen activation and reveal higher activity in relation to peroxide radicals $RO_2^{\cdot}$ in comparison with $Ni(II)(L^1)_2 \cdot 18K6_n$ (L^1=$(enamac)^-$ or $(acac)^-$, n=1,2). Values of w_{pr}^0 are in ~ 3 times higher than w_i^0. Decomposition of PEH makes significant contribution into formation of AP and MPC in developed reaction [51].

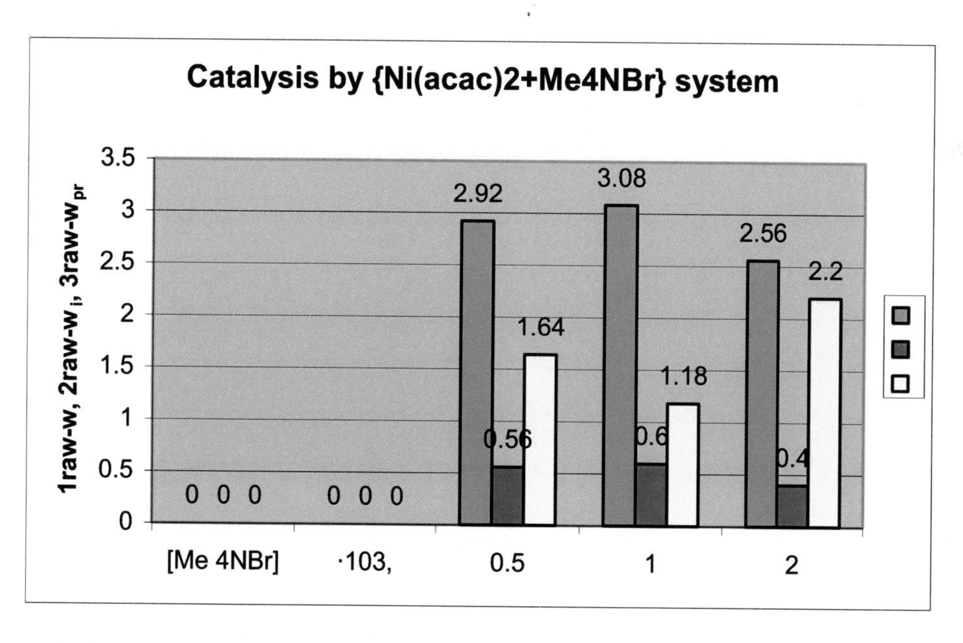

Figure 4. The rates of ethylbenzene oxidation products accumulation at the beginning of reaction ($w \cdot 10^5$) and calculated rates of chain initiation ($w_i \cdot 10^6$) and propagation ($w_{pr} \cdot 10^6$) ($M \cdot sec^{-1}$). Catalysis by system $\{Ni(II)(acac)_2 + Me_4NBr\}$. $[Ni(acac)_2]$=1,5·10^{-4} M, 120°C.

In [29] they show that formation of catalyst responsible for selectivity of ethylbenzene oxidation in the presence of catalytic system {$Co(acac)_2$ + HPMA} is not connected with oxidation of ligand $(acac)^-$ by molecular oxygen. Similarity of phenomenology of ethylbenzene oxidation catalyzed by systems {$Co(acac)_2$ + HMPA} and {$Co(acac)_2$ + 18K6} allow assuming [33] that transformation of complexes $Co(acac)_2 \cdot 18K6_2$ occurs under the action of radicals $RO_2^{\cdot}$ with formation of new catalytic particles inactive in PEH decomposition. Complexes $[Co(III)(L^1)_2 \cdot L^{2} \cdot (RO_2^-)]$ [60] may be such particles. We [29] and also authors of [18] established that in the absence of L^2 the transformation of $Co(acac)_2$ in the course of oxidation at [Cat] $\leq 1,5 \cdot 10^{-4}$ M occurs by mechanism of ligand $(acac)^-$ oxidation.

By the value of parameter $\tilde{S} \cdot C$ complexes of cobalt salts with macrocycle polyether 18K6 (2) are more active catalysts of ethylbenzene oxidation into PEH in comparison with complexes $Co(acac)_2 \cdot 18K6_2$ (Table 1). Rise of S_{PEH} and reduction of rate of side products AP and MPC accumulation in the course of oxidation reaction in the presence of $2Co(NO_3)_2 \cdot 18K6 \cdot 6H_2O$ (2) testify to formation of more active in selective catalysis particles. Transformation of complexes (2) (in the absence of L^1) probably occurs by the mechanism analogous to oxidation of complexes $Co(acac)_2$ with L^2 (L^2=HPMA, 18K6) as a result of interaction with peroxide radicals [60]. Estimation of values w_{pr} and w_i (formulas 1 and 2) showed that in given case the reaction of chain propagation Cat+$RO_2^{\cdot} \rightarrow$ gained in significant importance in catalysis mechanism. $w_{pr} > w_i$ at the beginning of reaction and in developed reaction (in 3,3 and 6,6 times correspondingly).

The rise of S_{PEH}^0 from 50 up to ~ 65-70% in reaction of ethylbenzene oxidation catalyzed by system {$Fe(III)(acac)_2$+18K6} ($[18K6]^0$=$0,5 \cdot 10^{-3}$ M, $5,0 \cdot 10^{-3}$ M) is connected with decrase of rates of AP and MPC formation at stages of chain propagation Cat + $RO_2^{\bullet} \rightarrow$. w_{pr}^0 are decreased in 1,6 and 3,5 times correspondingly. At that w_i^0 is either constant ([18K6]= $0,5 \cdot 10^{-3}$ M), or is reduced in 3,3 times ([18K6]= $5 \cdot 10^{-3}$ M) [55].

3. Catalysis by Triple Catalytic Systems Including Ni(II)(Acac)$_2$, Electro-Donor Monodentate Ligand L^2 (L^2= MSt (M=, Na, Li), N-Methylpyrrolidone-2, HMPA) and Phenol

One of the most effective methods of selectivity and conversion degree increase of ethylbenzene oxidation into α-phenylethylhydroperoxide (PEH) by molecular oxygen under catalysis by $M(L^1)_n$ may be application together with additives of electron-donor ligands L^2 of the third component of catalytic system – phenol (P.), the product of ethylbenzene oxidation, with formation of it in reaction sharp decrease of selectivity of ethylbenzene into PEH oxidation is connected. We established that efficiency of catalytic system {$M(L^1)_2$ + L^2} (M=Ni(II), L^1=$acac^-$, L^2= MSt (M=, Na, Li) N-methylpyrrolidone-2, HMPA) estimated by the value of parameter $\tilde{S} \cdot C$ was significantly increased in the presence of phenol (P.).

While investigating dependence of parameter $\tilde{S} \cdot C$ on $[L^2]$ in oxidation reaction in the presence of {$Ni(II)(acac)_2$ + L^2 + P.} (L^2=MP) at $[Ni(II)(acac)_2]$=const=$3 \cdot 10^{-3}$ M and [P.]=const=$3 \cdot 10^{-3}$ M (120°C) it turned out that dependence is extreme. (Here S is also averaged selectivity under change of S_{PEH} in the limits $S_0 < S \leq 80\%$, C – conversion degree as S_{PEH}=80%). Maximum value of $\tilde{S}\cdot C$ is reached at [MP]= $7\cdot10^{-2}$ M (S^{max}=85-87%). Concentration [MP]= $7\cdot10^{-2}$ M corresponds to formation of complexes of $Ni(II)(acac)_2$ with MP of structure 1 : 1 (in the absence of P.) [29]. It is characteristic that the value $(\tilde{S}\cdot C)^{max}$ $=17,5\cdot10^2$ (%,%) exceeds value $\tilde{S}\cdot C$ for complexes $Ni(II)(acac)_2\cdot MP$ ($11,9\cdot10^2$ (%,%)) and coordinated saturated complexes $Ni(II)(acac)_2\cdot 2MP$. Observing significant synergetic effect of parameter $\tilde{S}\cdot C$ increase under catalysis by $\{Ni(II)(acac)_2 + L^2\}$ in the presence of inhibitor phenol may be explained by unusual catalytic activity of formed at mentioned conditions triple complexes $[M(L^1)_2\cdot(L^2)_n\cdot(P.)_m]$. This presumption is confirmed by dependences of $\tilde{S}\cdot C$ on $[Ni(II)(acac)_2]$ at [P]=const=$3\cdot10^{-3}$ M and [MP]=const=$7\cdot10^{-2}$ M ($(\tilde{S}\cdot C)^{max}=17,47\cdot10^2$ (%,%), $[Ni(II)(acac)_2]=3\cdot10^{-3}$ M) and also of $\tilde{S}\cdot C$ on [P.] at $[Ni(II)(acac)_2]$=const=$3\cdot10^{-3}$ M and [MP]=const=$7\cdot10^{-2}$ M. In last case $\tilde{S}\cdot C$ reaches extremum $(\tilde{S}\cdot C)^{max}$=17,5 and $18,12\cdot10^2$ (%,%) at two values of [P.] = $3\cdot10^{-3}$ и $4,6\cdot10^{-4}$ M accordingly (Figure 5).

Comparison of kinetic regularities of ethylbenzene oxidation catalyzed by triple systems $\{Ni(II)(acac)_2(3\cdot10^{-3}$ M) + MP($7\cdot10^{-2}$ M) + P.} ([P.]= $3\cdot10^{-3}$ or $4,6\cdot10^{-4}$ M) was carried out. obtained data testify on the fact that in both cases selective catalysis of ethylbenzene oxidation into PEH is connected with formation in the course of oxidation of catalytically active complexes with structure 1 : 1 : 1 [unpublished data].

Analogous increase of efficiency of catalysis of ethylbenzene selective oxidation into PEH under introduction into system of third component phenol was also observed for a number of other binary systems $\{Ni(II)(acac)_2 + L^2\}$ (L^2 = NaSt, LiSt, HMPA) (Table 3).

Phenomenal results were obtained in the case of $\{Ni(II)(acac)_2$ ($3,0\cdot10^{-3}$ M) + NaSt ($3,0\cdot10^{-3}$ M) + P.($3,0\cdot10^{-3}$ M)}. Selectivity of oxidation into PEH $S_{PEH} = S_{PEH}^{max}$ =85-87% remains to significant depth of reaction C > 35%, parameter $\tilde{S}\cdot C \sim 30,1\cdot10^2$ (%,%) exceeds $\tilde{S}\cdot C = 24,3\cdot10^2$ (%,%) for $\{Ni(II)(acac)_2 + Me_4NBr\}$ (Table 1). Concentration $[PEH]^{max}$ = 1,6 -1,8 M (~27 mass %), that is higher than $[PEH]^{max}$ for all studied earlier binary catalytic systems $\{Ni(II)(acac)_2 + L^2\}$ and also the most effective triple catalytic systems $\{Ni(II)(acac)_2 + L^2 + P.\}$ (Table 3).

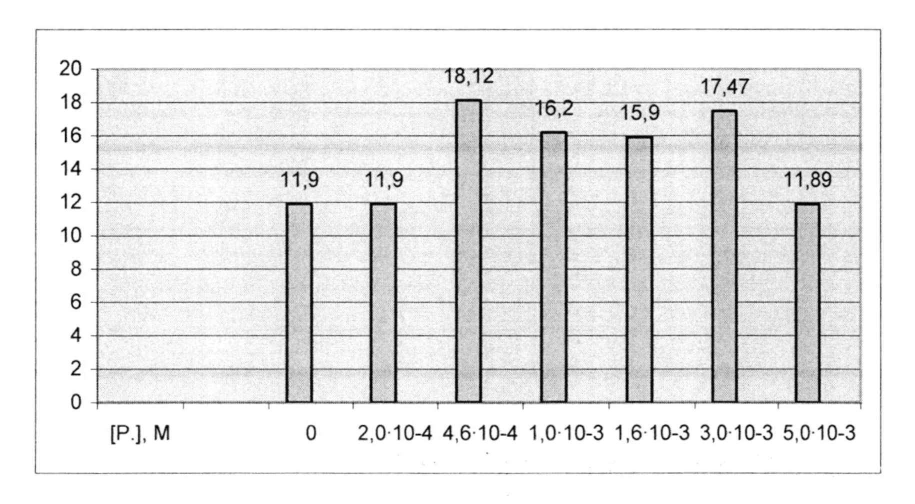

Figure 5. Dependence of $\tilde{S}\cdot C\cdot10^{-2}$(%,%) on [P.] in reaction of ethylbenzene oxidation catalyzed by $\{Ni(II)(acac)_2$ + MP + P.}. $[Ni(II)(acac)_2]$= const=$3\cdot10^{-3}$ M, [MP]=const=$7\cdot10^{-2}$ M, 120°C.

The data presented in Table 3 are protected by patent RU No.2002133752. Registration date is 11.2.2004; the authors are L.I. Matienko, L.A. Mosolova, patent holder is N.M. Emanuel's Institute of biochemical physics RAS.

Similarity of phenomenology of ethylbenzene oxidation in the presence of {$Ni(II)(acac)_2$ ($3{,}0\cdot10^{-3}$ M) + NaSt ($3{,}0\cdot10^{-3}$ M) + P.($3{,}0\cdot10^{-3}$ M)} and {$Ni(II)(acac)_2$ ($3{,}0\cdot10^{-3}$ M) + MP ($7{,}0\cdot10^{-2}$ M) + P.($3{,}0\cdot10^{-3}$ M)} allows assuming analogous mechanism of selective catalysis realizing by formed in the course of oxidation triple complexes. Parallel formation of PEH and side products AP and MPC is established: $w_P/w_{PEH} \neq 0$ at $t\to0$ (P=AP or MPC) and $w_{AP}/w_{MPC}\neq0$ at $t\to0$ at the beginning of reaction and in developed reaction of ethylbenzene oxidation catalyzed by {$Ni(II)(acac)_2$+L^2+P.} (L^2 = NaSt, MP). Increase of S_{PEH} in developed reaction under catalysis by complexes $Ni(II)(acac)_2\cdot L^2\cdot$P., as well as in the case of catalysis by $Ni(II)_x(L^1)_y(L^1{}_{ox})_z\cdot L^2$ in comparison with non catalyzed oxidation is connected with the change of PEH, AP and MPC formation direction and also with hindering of chain and heterolytic decomposition of PEH.

Increase of parameter $\tilde{S}\cdot C$ in the presence of {$Ni(II)(acac)_2$+L^2+P.} in comparison with catalysis by binary system {$Ni(II)(acac)_2$+L^2} is obviously conditioned by higher stability of formed in the process of oxidation active catalytic particles responsible for selectivity of ethylbenzene oxidation into PEH. Complexes $Ni(II)(acac)_2\cdot L^2\cdot$P. Are active for long time, and ligand $(acac)^-$ is not transformed in the course of reaction. In this connection obviously, introduction into reaction mixture {$Ni(II)(acac)_2$+L^2} of the third component phenol is one of the most effective methods of optimization of ehtylbenzene oxidation into PEH catalyzed by {$Ni(II)(acac)_2$+L^2}.

The mechanism of catalysis of ethylbenzene oxidation by nickel complexes in the presence of additives of electron-donor monodentate ligands established by us allows solving of problem of selective oxidation into ROOH using in particular the following methods of selective oxidation process optimization:

(1) introduction of activated additives in the course of oxidation process;
(2) application of macrocycle polyethers or quaternary ammonium salts as extra ligands-modifiers.

Moreover, the method of selective oxidation optimization was suggested consisting in introduction into reaction system of the third component of catalytic system, one of the products of reaction, i.e. phenol. Reduction of selectivity of oxidation into PEH is connected with formation of phenol in the course of reaction catalyzed by {$Ni(II)(L^1)_2$+L^2}. Mechanism of activation of catalytic system {$Ni(II)(L^1)_2$+L^2} is consisted in regeneration of catalyst active form, multi-ligand complex "A"; increase of stationary concentration of "A" (1); in increase of activity of primary complexes $Ni(II)(L^1)_2\cdot(L^2)_n$; stabilization of formed in the course of oxidation complexes of "A" type $Ni(II)_x(L^1)_y(L^1{}_{ox})_z\cdot L^2$ (2); in formation of more stable in comparison with "A" complexes catalytically active triple complexes $M(L^1)_2\cdot L^2\cdot$P.(3).

Estimation of catalytic activity of nickel, cobalt and iron complexes with 18K6 or R_4NBr at elementary stages of ethylbenzene oxidation process in the framework of radical-chain mechanism showed that rate and selectivity of ethylbenzene oxidation into PEH by initial complexes and formed in the course of oxidation process catalytic particles depended on the ratio between rates of chain initiation (w_i, activation O_2) and propagation (w_{pr}, Cat+$RO_2^{\bullet}\rightarrow$), w_{pr}/w_i, and also on activity of Cat in PEH decomposition (chain, homolytic, heterolytic), direction of products formation methylphenylcarbinole and acetophenone (parallel with PEH or consecutively under hydroperoxide decomposition).

Results of our investigation of ethylbenzene oxidation into PEH catalyzed by systems on the base of $Ni(II)(acac)_2$ and R_4NBr were used by authors of [61, 62] for developing of catalysts for industrial oxidation of ethylbenzene into PEH at the Repsol factory of Puertollano (Spain) for joint production of propylene oxide and styrene. Tetrafluorine borate of tetra-n-butylammonium and hexafluorine phosphate of 1-n-butyl-3-methyl imidazoly were used as onium salts. At that **S** = 90-95 - 80% was observed at deep stages of oxidation not exceeding **C** = 10-12%, that was significantly lower than in the case of catalysis by system {$Ni(II)(acac)_2$ + Me_4NBr}. Authors referred on our works and used method for estimation of catalytic activity of catalysts at elementary stages of chain initiation and propagation suggested by us.

References

[1] E.T. Denisov, N.M. Emanuel, *Uspekhi khimii*, 29, 1409 (1960) *(in Russian).*

[2] N.M. Emanuel, E.T. Denisov, Z.K. Maizus, *Chain reactions of hydrocarbons oxidation in liquid phase*, Moscow: Nauka (1965) *(in Russian).*

[3] I.V. Berezin, E.T. Denisov, N.M. Emanuel, *Cyclohexane oxidation*, Moscow: MSU (1962) *(in Russian).*

[4] L.I. Matienko, L.A. Goldina, I.P. Skibida, Z.K. Maizus, *Izv. AN SSSR, Ser. Khim.*, No.2, 287 (1975) *(in Russian).*

[5] L.I. Matienko, I.P. Skibida, Z.K. Maizus, *Izv. AN SSSR, Ser. Khim.*, No.6, 1322 (1975) *(in Russian).*

[6] L.I. Matienko, *Dissertation of candidate of science*, Moscow: Institute of chemical physics AN SSSR (1976) *(in Russian).*

[7] I.P. Skibida, *Uspekhi khimii*, 44, No.10, 1729 (1975) *(in Russian).*

[8] R.A. Sheldon, J.K. Kochi, In: Advances in Catalysis / Ed. by Eley D.D., Pines H., Weiz P.B. New York, San Francisco, London: Acad. Press (1976).

[9] R.A. Sheldon, J.K. Kochi, *Metal-Catalyzed Oxidation of Organic Compounds*, New York, London: Acad. Press (1981).

[10] R.A. Sheldon, *J. Mol. Catal.*, 20, 1 (1983).

[11] H. Mimoun, *Chem. and Phys. Aspects of Catal. Oxid.*, Paris: CNRS (1981).

[12] R.A. Sheldon, *The Activation of Dioxygen and Homogeneous Catalytic Oxidation*, Ed. by Barton D.H.R., Martell A.E., Sawyer D.T. New York: Plenum Press, 1993.

[13] T. Mlodnicka, *Metalloporphyrins in Catalytic Oxidation*, Ed. Sheldon R.A. New York, Basel, Hong Kong: Marcel Dekker, Inc. (1994).

[14] N.M. Emanuel, *Uspekhi khimii*, 47, No.8, 1329 (1978) *(in Russian).*

[15] W. Partenheimer, *Catalysis Today*, 23, 69 (1991).

[16] D. Mansuy, *The Activation of Dioxygen and Homogeneous Catalytic Oxidation*, Eds. Barton D.H.R., Martell A.E., Sawyer D.T. New York: Plenum Press (1993).

[17] K. Weissermel, H.-J. Arpe, *Industrial Organic Chemistry*, 3nd ed., transl. by C.R. Lindley. New York: VCH (1997).
[18] N.M. Emanuel, D.Gal, *Ethylbenzene oxidation. Model reaction*, Moscow: Nauka (1984) *(in Russian)*.
[19] Yu.D. Norikov, E.A. Blyumberg, L.V. Salukvadze, *Problemy kinetiki i kataliza*, No.16, Moscow: Nauka (1975) *(in Russian)*.
[20] M.V. Nesterov, V.A. Ivanov, V.M. Potekhin, V.A. Proskuryakov, M.Yu. Lysukhin, *Zh. Prikl. Khimii*, 52, No.7, 1589 (1979) *(in Russian)*.
[21] L.A. Mosolova, L.I. Matienko, Z.K. Maizus, *Izv. AN SSSR, Ser. Khim.*, No.8, 1760 (1978) *(in Russian)*.
[22] L.A. Mosolova, L.I. Matienko, Z.K. Maizus, *Izv. AN SSSR, Ser. Khim.*, No.2, 278 (1980) *(in Russian)*.
[23] L.A. Mosolova, L.I. Matienko, Z.K. Maizus, *Izv. AN SSSR, Ser. Khim.*, No.4, 731 (1981) *(in Russian)*.
[24] L.A. Mosolova, L.I. Matienko, Z.K. Maizus, *Izv. AN SSSR, Ser. Khim.*, No.9, 1977 (1981) *(in Russian)*.
[25] L.A. Mosolova, L.I. Matienko, *Neftekhimiya*, 25, No.4, 540 (1985) *(in Russian)*.
[26] L.A. Mosolova, L.I. Matienko, I.P. Skibida, *Kinetika i kataliz*, 29, No.5, 1078 (1988) *(in Russian)*.
[27] L.I. Matienko, Z.K. Maizus, *Kinetika i kataliz*, 15, No.2, 317 (1974) *(in Russian)*.
[28] L.A. Mosolova, L.I. Matienko, Z.K. Maizus, E.M. Brin, *Kinetika i kataliz*, 21, No.3, 657 (1980) *(in Russian)*.
[29] L.A. Mosolova, L.I. Matienko, I.P. Skibida, *Kinetika i kataliz*, 28, No.2, 479 (1987) *(in Russian)*.
[30] L.I. Matienko, L.A. Mosolova, *Izv. AN SSSR, Ser. Khim.*, No.1, 55 (1999) *(in Russian)*.
[31] M. Utaka, M. Hojo, Y. Fujiu, A.. Takeda, *Chem. Lett.*, 635 (1984).
[32] T. Sagawa, K. Ohkibo, *J. Mol. Catal. A: Chem.*, 113, 269 (1996).
[33] L.A. Mosolova, L.I. Matienko, I.P. Skibida, *Kinetika i kataliz*, 31, No.6, 1377 (1990) *(in Russian)*.
[34] L.A. Mosolova, L.I. Matienko, I.P. Skibida, *Izv. AN SSSR, Ser. Khim.*, No.8, 1406 (1994) *(in Russian)*.
[35] L.I. Matienko, L.A. Mosolova, *Kinetika i kataliz*, 44, No.2, 237 (2003) *(in Russian)*.
[36] N.M. Emanuel, In coll.: *Questions on chemical kinetics, catalysis and reaction ability*, Moscow: AN SSSR (1955) *(in Russian)*.
[37] P. Bonchev, *Complex formation and catalytic activity*, Moscow: Mir (1975) *(in Russian)*.
[38] M. Basato, B. Corain, P. De Roni, G. Favero, R. Jaforte, *J. Mol. Catal.*, 42, No.1, 115 (1987).
[39] G.A. Kovtun, D.L. Lysenko, I.I. Moiseev, In coll.: *XII Mendeleev conference on general and applied chemistry*, Moscow: Nauka, No.3, 45 (1981) *(in Russian)*.
[40] V.I. Vedeneev, E.A. Frankevich, V.N. Kondrat'ev, V.A. Medvedev, *Energy of chemical bonds breakage. Ionization potentials and affinity to electron*, Moscow: AN SSSR (1962) *(in Russian)*.
[41] R. Foster, *Organic – Transfer Complexes*, N.Y. – L.: Acad. Press (1969).
[42] V.E. Kampar, O.Ya. Neiland, *Uspekhi khimii*, 46, No.6, 945 (1977) *(in Russian)*.

[43] R. Cammac, In: *Adv. Inorg. Chem.*, Ed. by Sykes A.G. N.Y.-L.-Tokyo: Acad. Press Inc., 32, 297 (1989).
[44] M.A. Halcrow, G. Chistou, *Chem. Rev.*, 94, 2423 (1994).
[45] V.K. Bel'skii, B.M. Bulychev, *Uspekhi khimii*, 68, No.2, 136 (1999) *(in Russian).*
[46] R.N. Mc Donald, A.K. Chowdhury, W.Y. Gung, K.D. De Witt, In: *Nucleophility. Adv. Chem. Ser.*, Ed. by Harris J.M., Mc Manus S.P. Washington: Am. Chem. Soc. (1987).
[47] E.V. Demlov, *Izv. AN SSSR, Ser. Khim.*, No.11, 2094 (1995) *(in Russian).*
[48] Ms.M. Satpathy, B. Pradhan, *Asian J. Chem.*, 9, No.4, 873 (1997).
[49] H. Yamaguchi., K. Katsumata, M. Steiner, H.J. Miketa, *J. Magn. Magn. Mater.*, 177-181(Pt.1), 750 (1998).
[50] J.H. Clark, J.M. Miller, *J. Chem. Soc. Perkin Trans.*, No.1, 1743 (1977).
[51] L.I. Matienko, L.A. Mosolova, *Izv. AN SSSR, Ser. Khim.*, No.4, 689 (1997) *(in Russian).*
[52] T.V. Maximova, T.V. Sirota, E.V. Koverzanova, O.T. Kasaikina, *Neftekhimiya*, 41, No.5, 289 (2001) *(in Russian).*
[53] V.A. Golodov, *Ross. Khim. Zh.*, 44, No.3, 45 (2000) *(in Russian).*
[54] L.I. Matienko, L.A. Mosolova, *Kinetika i kataliz*, 46, (2005) in prints *(in Russian).*
[55] L.I. Matienko, L.A. Mosolova, In coll.: *Organic catalysis and heterocycles*, Moscow: Khimiya (in prints) *(in Russian).*
[56] R.D.Arasasingham, Ch.R. Cornman, A.L. Balch, *J. Am. Chem. Soc.*, 111, No.20, 7800 (1989).
[57] F.A. Chaves, J.M. Rowland, M.M. Olmstead, P.K. Mascharak, *J. Am. Chem. Soc.*, 120, No.22, 9015 (1998).
[58] E. Solomon-Rapaport, A. Masarwa, H. Cohen, D. Meyerstein, *Inorg. Chim. Acta.*, No.299, 41 (2000).
[59] A.E. Semenchenko, V.M. Solyanikov, E.T. Denisov, *Zh. Phiz. Khimii*, 47, No.5, 1148 (1973) *(in Russian).*
[60] E.P. Talzi, V.D. Chinakov, V.P. Babenko, V.N. Sidelnikov, K.I. Zamaraev, *J. Mol. Catal.*, 81, 215 (1993).
[61] R. Alcantara, L. Canoira, P. Guilherme-Joao, J.-M. Santos, I. Vazquez, *Appl. Catal. A: Gener.*, 203, 259 (2000).
[62] R. Alcantara, L. Canoira, P. Guilherme-Joao, P. Perez-Mendo, *Appl. Catal. A: Gener.*, 218, 269 (2001).

In: Reactions and Properties of Monomers and Polymers ISBN: 1-60021-415-0
Editors: A. D'Amore and G. Zaikov, pp. 43-67

Chapter 3

LIQUID-PHASE OXIDATION OF ALKANES IN THE PRESENCE OF METALS COMPOUNDS

E. I. Karasevich
N.M. Emanuel's Institute of Biochemical Physics Russian Academy of Sciences
4, Kosygin str., Moscow 111999, Russia

ABSTRACT

The review of investigations of alkanes oxidation by molecular oxygen in the presence of metals compounds in organic solvents is presented. The examples of systems of radical-chain and biomimetic oxidation of alkanes are presented. Kinetics and mechanisms of active particles and hydrocarbons oxidation products formation in these systems are considered.

Keywords: catalysis, kinetics, oxidation, molecular oxygen, biomimetism, metal-complexes, iron porphyrins, alkanes, cyclohexane.

INTRODUCTION

Reaction of chemical substances oxidation by molecular oxygen is one of the most important among all chemical processes. As a result of reactions of deep oxidation – breathing and combustion – the energy stored in organic substances by photosynthesis is liberated. Process of hydrocarbons slow oxidation fills a highly important place in petrochemistry as very effective method of reception of many valuable chemical products. Liquid-phase oxidation of hydrocarbons proceeds at lower temperatures, more "softly" in comparison with gas-phase oxidation of the same substances [1, 2].

At the end of the century before last Bakh and Engler proposed peroxide theory of slow oxidation processes the main position of which was consisted in the fact that primary products of organic substances oxidation by molecular oxygen were peroxide compounds in which two oxygen atoms remained bonded with each other [3]. The fact about peroxides formation took on special significance in Professor N.N. Semenov's theory of degenerated-branched reactions [4, 5] and later developed in N.M. Emanuel's works for liquid-phase process of hydrocarbons oxidation [1, 2, 6]. Peroxides turned out to be responsible for chain branching leading to oxidation process self-acceleration. For each concrete hydrocarbon there are its own particularities, however degenerated-branched mechanism of non-catalyzed liquid-phase oxidation of hydrocarbons by molecular oxygen has general meaning. As a result of numerous studies carried out by Professor N.M. Emanuel the principle scheme of oxidation including stages of chain initiation, branching, propagation and termination was established [6, 7].

Stimulating influence of various metals compounds on slow chain branched reactions is old-established fact. Transition metals salts are widely used while realizing of many technological processes of hydrocarbon oxidation. Metals ions may participate in all stages of chain process of oxidation: chain initiation, propagation and termination [1, 8]. Variety of mechanisms of catalysis by metals compounds is explained by the last fact.

Functionalization of alkanes fills special place among hydrocarbons oxidation processes. The main sources of alkanes are oil and natural gas that are used mainly as fuel. Only 5% of milliards of tons of produced oil are chemically processed and direct functionalization of alkanes is equal to only parts of percent. The reason of such situation is in chemical inertness of alkanes reflected in their old name "paraffin" (from Latin "parum affinis" that means "lacking in affinity"). At the beginning of sixties of the last century N.M. Emanuel with co-authors published the work [9] summarizing results of investigation of questions on kinetics and chemism of important for petrochemistry reaction of oxidation of saturated hydrocarbon cyclohexane. Cyclohexane oxidation reaction is presented in this work as model one for processes of hydrocarbons liquid-phase oxidation and is considered in wide aspect – from elementary reactions to principles of process technological design. Analogously to a lot of other radical-chain processes of hydrocarbons oxidation cyclohexane oxidation leading to formation of cyclohexanol, cyclohexanone and adipinic acid, i.e. raw materials for production of synthetic fibers requires high temperatures (100-200°C) and pressures (10-50 atm) [9].

For increase of oxidation processes selectivity and efficiency expensive and ecologically insecure oxidizers such as nitric acid, perchlorates, hypochlorites are widely used in technological processes as oxidizers. These oxidizers as a rule are expensive enough and that is more important their application leads to formation of nonutilizable waste. Creation of novel catalytic systems of selective oxidation of organic compounds with the use of more ecologically pure and cheap oxidizer – molecular oxygen is a problem of current importance of chemical science and practice to which N.M. Emanuel attended great attention [6]. Significant progress in this field is outlined recently; however mankind practical needs strongly require the search of novel alkanes reactions for creation of effective processes of their transformation. At the same time, in animate nature the problem of organic compounds oxidation including alkanes is successfully solved. Corresponding ferment systems are still unsurpassed in relation to both rate and selectivity of alkanes oxidation, that is why usage of animate nature functioning principles, i.e. biomimetic approach to creation of chemical technologies seems to be very perspective and first of all from ecological point of view [10]. At present investigations in the field of hydrocarbons liquid-phase oxidation are developed in two directions. From the one hand, they search for selective catalytic systems of radical-chain oxidation of organic compounds. Furthermore, principally novel systems are constructed; principles of nature systems work – monooxygenase – are meaningly used for their development.

Results of works series carried out for continuation of N.N. Semenov's and N.M. Emanuel's works are presented in given review. Early works of this series were repeatedly discussed by N.M. Emanuel in his reviews devoted to problems of liquid-phase oxidation of hydrocarbons.

1. CONJUGATED OXIDATION OF ALKANES AND METALS SALTS BY MOLECULAR OXYGEN

In seventies of the last century the systems were proposed that worked by principle of oxidation conjugated with auto-oxidation of metals chlorides (Fe, Sn, V, Ti, Mo, W) and oxidized hydrocarbons (saturated, unsaturated and aromatic) with high selectivity at room temperature [11]. Distinguishing feature of these systems is application of reducer (inductor) in oxidation process for generation of active particles able to attack the strongest C–H bonds. In these systems oxidation of hydrocarbons even such inert as methane and ethane occurs at soft conditions. Hydrocarbon skeleton of oxidized molecules remains, limited number of products is formed (Table 1).

At first glance, it is hard to correlate high selectivity by product observed in these systems with radical mechanism. However, soft conditions limit the number of possible reactions increasing selectivity of process. Candidates to active reagents directly interacting with hydrocarbons in these systems are radicals $^{\bullet}OH$ and $HO^{\bullet}_2$ and also metals oxo-complexes. Experiments showed [11], that rate of cyclohexane oxidation is reduced with increase of rate constant of radicals $^{\bullet}OH$ interaction with organic compounds used as solvents in these systems.

Table 1. Conjugated oxidation of cyclohexane and metals chlorides in water-acetonitrile solutions. Conditions are: 0,1 MPa of air, $[H_2O]$=1M, [HCl]=0,1M, $[MCl_n]_0$=0,04M, $[C_6H_{12}]$=0,23M, temperature 25°C.

M	Change of metal oxidation degree as a result of reaction	Reaction duration, min	$[C_6H_{10}O]\cdot10^4$, M	$[C_6H_{11}OH]\cdot10^4$, M	Yield of oxidation products C_6H_{12}, per MCl_n, %
Sn	II → IV	1	<0,1	22	5
Fe	II → III	3	6,5	30	9
Ti	III → IV	15	1,5	32	8
V	II → IV	60	<0,1	30	7
Mo	III → V	420	<0,1	16	4
W	III → VI	300	2,6	45	12

This fact testifies in favour of hydrocarbons oxidation mechanism the first stage of which is interaction with $^{\bullet}OH$ able to remove H atom at 20°C even from the most inert methane. Profs for such mechanism were obtained under detailed investigation of kinetics of conjugated oxidation of cyclohexane and auto-oxidation of divalent tin oxide in water-organic solutions. Selection of $SnCl_2$ later turned to be successful due to variety of kinetic effects in this system. Cyclohexane was selected as model substrate under the influence of works by Professor N.M. Emanuel [9]

1.1. Kinetics and Mechanism of $SnCl_2$ Auto-Oxidation in Water-Organic Solutions

Kinetics of oxidation of crystalline hydrate tin (II) chloride in water-organic solutions was studied by volumetric method by oxygen absorption [12-14]. Water concentration in solution significantly influences on process rate and kinetic curves character. It is convenient to select three regions:

(1) water-organic solutions in which $[H_2O] > [SnCl_2]$;
(2) waterless solutions in which $[H_2O] \leq [SnCl_2]$ all added water enters the coordination sphere of tin;
(3) water solutions.

In water-organic solutions at $[H_2O]$ = (1-3)M process of xodation by air oxygen proceeds without noticeable induction period, rate of O_2 absorption is constant almost up to the end of reaction (Figure 1).

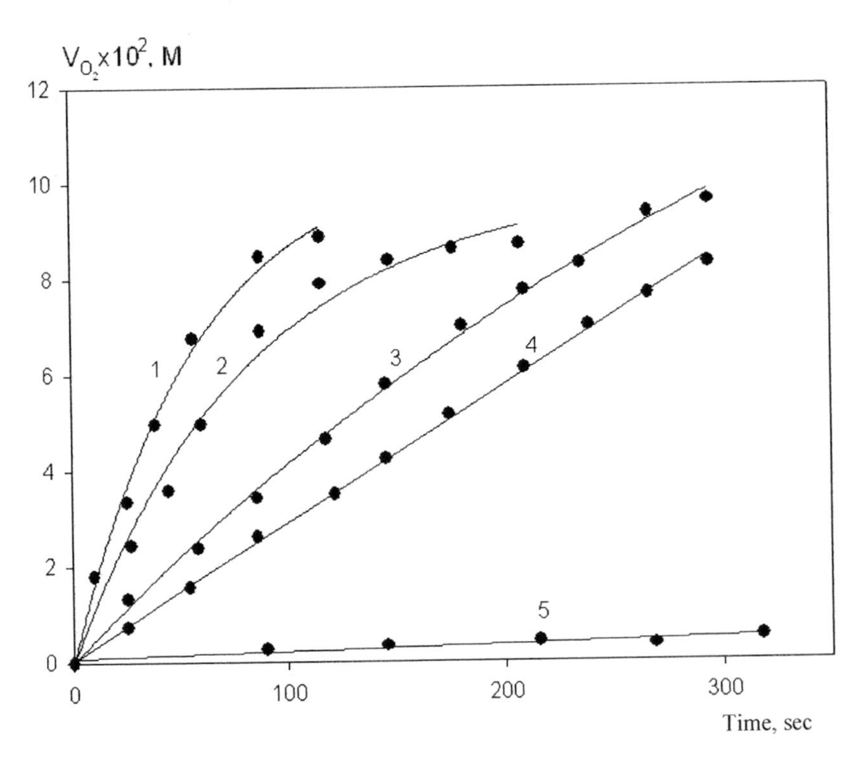

Figure 1. Kinetic curves of oxygen absorption under auto-oxidation of 0,175M $SnCl_2 \cdot 2H_2O$ in water-organic solvents: 1 – acetonitrile, 2 – acetone, 3 – methanol, 4 – tret-butanol, 5 – dimethylformamide. Conditions are: 0,1 MPa of air, $[H_2O]$=2M, temperature 25°C.

At first glance such simple kinetics looks like non-branched chain or absolutely non-chain process. However with decrease of oxygen concentration kinetic curves take on auto-catalytic character. Induction periods appear, and they are the bigger, the smaller concentration of O_2. Kinetic law for oxygen absorption rate in developed process of $SnCl_2$ auto-oxidation is changed when water concentration is changed. In the whole interval of water concentration change introduction of inhibitors of radical-chain oxidation into system leads to appearance or increase of induction period. Dependences of induction period duration on inhibitors concentration have critical character (Figure 2). The presence of critical phenomena is impossible to explain by mechanism with straight chains suggested earlier for auto-oxidation of divalent tin chloride in water mediums [15].

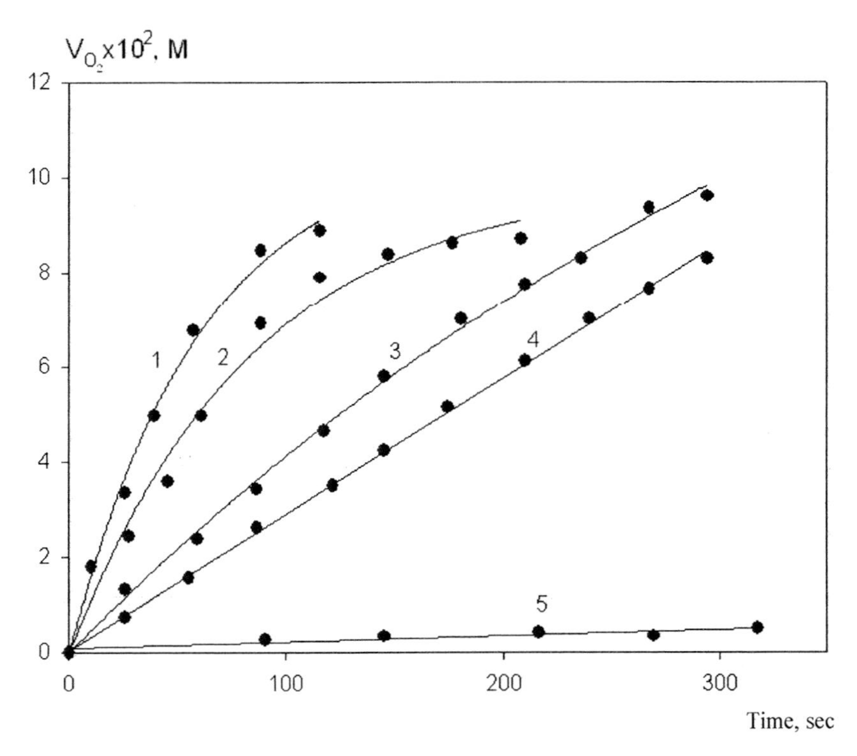

Figure 2. Dependence of induction period duration (τ_{ind}) of $SnCl_2 \cdot 2H_2O$ auto-oxidation in water-acetonitrile solutions on inhibitors concentration: 1 – α-naphthylamine, 2 – tetranitromethane, 3 – 1,2-naphthoquinone. Conditions are: 0,1 MPa of air, $[H_2O]$=(2-7)M, $[Sn(II)]_0$=0,2M, temperature 25°C.

On the base of analysis of obtained data we propose unified branched-chain mechanism of $SnCl_2$ oxidation in water, water-organic and water-less solutions which may be presented by scheme:

0. $Sn(II) + O_2 \xrightarrow{H_2O} Sn(III) + HO^{\bullet}_2$,

1. $Sn(III) + O_2 \longrightarrow Sn(IV)O^{\bullet -}_2$,

1'. $Sn(IV)O^{\bullet -}_2 \xrightarrow{H_2O} Sn(IV) + HO^{\bullet}_2$,

$$2.\ HO^{\bullet}_2 + Sn(II) \xrightarrow{H_2O} \begin{cases} \xrightarrow{\beta} H_2O_2 + Sn(III) \\ \xrightarrow{1-\beta} {}^{\bullet}OH + Sn(IV) \end{cases}$$

$$2'.\ Sn(IV)O^{\bullet -}_2 + Sn(II) \xrightarrow{H_2O} \begin{cases} \xrightarrow{\beta'} H_2O_2 + Sn(III) + Sn(IV) \\ \xrightarrow{1-\beta'} {}^{\bullet}OH + 2\ Sn(IV) \end{cases}$$

$$3.\ H_2O_2 + Sn(II) \longrightarrow \begin{cases} \xrightarrow{\alpha} {}^{\bullet}OH + Sn(III) \\ \xrightarrow{1-\alpha} H_2O + Sn(IV) \end{cases}$$

$4.\ H_2O_2 + Sn(III) \longrightarrow {}^{\bullet}OH + Sn(IV),$

$5.\ {}^{\bullet}OH + Sn(II) \longrightarrow Sn(III),$

$6.\ Sn(III) + Sn(III) \longrightarrow Sn(IV) + Sn(II),$

$7.\ Sn(IV)O^{\bullet -}_2 + Sn(III) \longrightarrow Sn(IV)\text{-}O\text{-}O\text{-}Sn(IV),$

$8.\ Sn(IV)O^{\bullet -}_2 + Sn(IV)O^{\bullet -}_2 \longrightarrow Sn(IV)\text{-}O\text{-}O\text{-}Sn(IV) + O_2,$

$9.\ Sn(IV)\text{-}O\text{-}O\text{-}Sn(IV) \xrightarrow{H_2O} H_2O_2 + 2Sn(IV),$

where α, β and β' – are the probabilities of one-electron way, 1-α, 1-β and 1-β' – probabilities of two-electron way.

Development of branched-chain process is connected with the fact that in reactions of chain propagation as chemical analysis shows [12] hydrogen peroxide is formed which under interaction with Sn(II) may give additional amount of Sn (II) being chain bearer.

Interaction of Sn(II) with H_2O_2 is realized mainly by two-electron way and probability of one-electron way leading to branching is equal to $8\cdot10^{-4}$ (H_2O), $5\cdot10^{-4}$ (MeCN), $1\cdot10^{-2}$ (acetone) [16]. However, branching of chain at the expense of this stage leads to accelerating accumulation of radical-like particles of Sn (III) at the beginning of oxidation process. Disproportionation of two particles of Sn (III) that is negative interaction of chains leads to stationary regime which is established after initial period of auto-accelerated reaction development.

Change of kinetics observed with water concentration change in solution is explained by the change pf chain termination character and H_2O_2 decomposition route. In water and water-organic solutions hydrolysis of $Sn(IV)O^{\bullet -}_2$ proceeds fast enough and the main stage of chain termination is reaction (6). In the case when hydrogen peroxide is decomposed mainly by reaction (3) the calculation by method of stationary concentrations in accordance with proposed scheme gives the following expression for oxidation rate:

$$W_1 = \alpha\beta(k_1^2/k_6)\,[O_2]^2 \qquad (I)$$

Such dependence is observed experimentally in water-organic solutions with $[H_2O]$ = (1-3) M [12].

In the case when hydrogen peroxide is destructed mainly by reaction (4) analogous calculation gives the following expression for oxidation rate:

$$W_2 = \left(\sqrt{(\alpha\beta \mathbf{k}_1^3 \mathbf{k}_3 / \mathbf{k}_4 \mathbf{k}_6)}\right) [SnCl_2]^{1/2} [O_2]^{3/2}, \qquad \text{(II)}$$

i.e. kinetic law observed in water solutions in the absence of organic solvents [15].

In the mixture of acetonitrile with water (1 : 1) the rate of auto-oxidation obeys the law intermediate between (I) and (II), i.e. hydrogen peroxide destruction is realized simultaneously by reactions (3) and (4). In this case not complicated calculation leads to the following expression of rate of oxygen absorption:

$$W = b[O_2][Sn(II)]\left\{\sqrt{1 + \frac{a[O_2]}{[Sn(II)]}} - 1\right\} \qquad \text{(III)}$$

where $\mathbf{a} = \dfrac{4\alpha\beta \mathbf{k}_1 \mathbf{k}_4}{\mathbf{k}_3 \mathbf{k}_6}$, $\mathbf{b} = \dfrac{\mathbf{k}_1 \mathbf{k}_3}{2\mathbf{k}_4}$.

It follows from expression (III) that:

$$\frac{\mathbf{k}_4[\mathbf{Sn(III)}]}{\mathbf{k}_3[\mathbf{Sn(II)}]} = \frac{1}{2}\left(\sqrt{(1 + \mathbf{a}\frac{[\mathbf{O}_2]}{[\mathbf{Sn(II)}]})} - 1\right), \qquad \text{(IV)}$$

i.e. at constant $[O_2]$ and [Sn(II)] ratio $\mathbf{k_4}$[Sn(III)]/$\mathbf{k_3}$[Sn(II)] is determined by the value of parameter **a**, which is the fuction of elementary constants.

Expression (III) may be presented as follows:

$$\frac{\mathbf{W}}{[\mathbf{Sn(II)}]\cdot[\mathbf{O}_2]} = -2\mathbf{b} + \mathbf{ab}^2 \frac{[\mathbf{O}_2]^2}{\mathbf{w}} \qquad \text{(V)}$$

Testing showed [14] that experimental points of dependences W on both [Sn(II)] and $[O_2]$ in various mediums lie down on lines in coordinates of equation (V) well, that confirms proposed mechanism of tin chloride oxidation. Increase of k_4[Sn(III)] under transition to water solutions may be caused by both increase of k_4 and [Sn(III)] at the expense of medium properties change.

Decrease of water concentration leads to increase of stationary concentration of $Sn(IV)O^{\bullet -}_2$ in relation to Sn(III) and necessity of consideration of termination reactions (7) and (8) in water-less solutions. According with made analysis hydrogen peroxide at these conditions is spent mainly by reaction (3). In this case the calculation by method of stationary concentrations gives the following expression for rate of oxygen absorption:

$$\mathbf{W} = \frac{\alpha\beta'\mathbf{k}_1\mathbf{k'}_2^2[\mathbf{Sn(II)}]^2[\mathbf{O}_2]}{\mathbf{k}_7\mathbf{k}_8(\frac{\mathbf{k}_1}{\mathbf{k}_7}[\mathbf{O}_2]+\frac{\mathbf{k'}_2}{\mathbf{k}_8}[\mathbf{Sn(II)}])} \quad \text{(VI)}$$

Such kinetic law is experimentally observed in water-acetone solutions at $[H_2O] \leq [SnCl_2]$ [14].

Thus, change of tin chloride auto-oxidation kinetics with the change of medium is well explained in the framework of unified branched-chain mechanism. Observed change of oxidation rate and changing of kinetic law are connected with the change of role of separate stages under water concentration variation. Serving as proton donor water accelerates process' elementary stages by conjugating of oxidation-reduction and acid-basic reactions (see reactions 1', 2 and 2'). At the same time, water while filling coordination sphere of tin hinders entering of reactive particles into it. Corresponding elementary stages at that are decelerated. This dual role of water – proton donor and ligand – explains in the whole the extreme character of oxidation rate dependence on $[H_2O]$.

1.2. Kinetics and Mechanism of Cyclohexane Oxidation Conjugated with Auto-Oxidation of Bivalent Tin Chloride

In the scheme of tin salt auto-oxidation considered above in details the most active particles are radicals $^{\bullet}OH$. Alkanes oxidation mechanism may be presented by consecution of stages analogous to stages of divalent tin chloride auto-oxidation:

10. $RH + {}^{\bullet}OH \longrightarrow R^{\bullet} + H_2O$

11. $R^{\bullet} + O_2 \longrightarrow RO^{\bullet}_2$

12. $RO^{\bullet}_2 + Sn(II) \xrightarrow{H_2O}$ $\xrightarrow{\beta''} ROOH + Sn(III)$; $\xrightarrow{1-\beta''} RO^{\bullet} + Sn(IV)$

13. $ROOH + Sn(II) \longrightarrow$ $\xrightarrow{\alpha'} RO^{\bullet} + Sn(III)$; $\xrightarrow{1-\alpha'} ROH + Sn(IV)$

14. $ROOH + Sn(III) \longrightarrow RO^{\bullet} + Sn(IV)$

15. $RO^{\bullet} + Sn(II) \longrightarrow Sn(III)$

16. $RO^{\bullet} + RH \longrightarrow R^{\bullet} + ROH$

17. $RO^{\bullet} + ROH \longrightarrow R'^{\bullet} + ROH$

In accordance with suggested scheme we may expect that alcohol will the only one product of C_6H_{12} oxidation that is really observed [11]. In classic liquid-phase radical-chain oxidation of cyclohexane [1] significant part of products is received by reactions $RO_2^\bullet$ + RH→molecular products; $RO_2^\bullet + RO_2^\bullet \rightarrow ROH + R'{=}O + O_2$. At conditions of conjugated oxidation these reactions practically don't proceed, radicals $RO_2^\bullet$ are spent by reaction (12) that leads to high selectivity of cyclohexane oxidation process.

For substantiation of concrete mechanism pf cyclohexane oxidation in water-acetonitrile solutions kinetic investigations were carried out [17]. They found that in studied system the rate of oxygen absorption is increased with the rise of C_6H_{12} concentration (Figure 3), and duration of induction period at that is reduced. At cyclohexane concentrations high enough induction period is not observed in reaction mixture, and rate of oxygen absorption W^{RH} is increased in more than three times in comparison with W^0 in the absence of cyclohexane. For non-branched mechanism of C_6H_{12} oxidation process, i.e. for $\alpha' = 0$, at maximum alcohol yield 30% per $[Sn(II)]_0$ experimentally observed at these conditions, we may expect the rise of oxygen absorption rate under introduction of cyclohexane into system of tin (II) auto-oxidation only in 1,3 times.

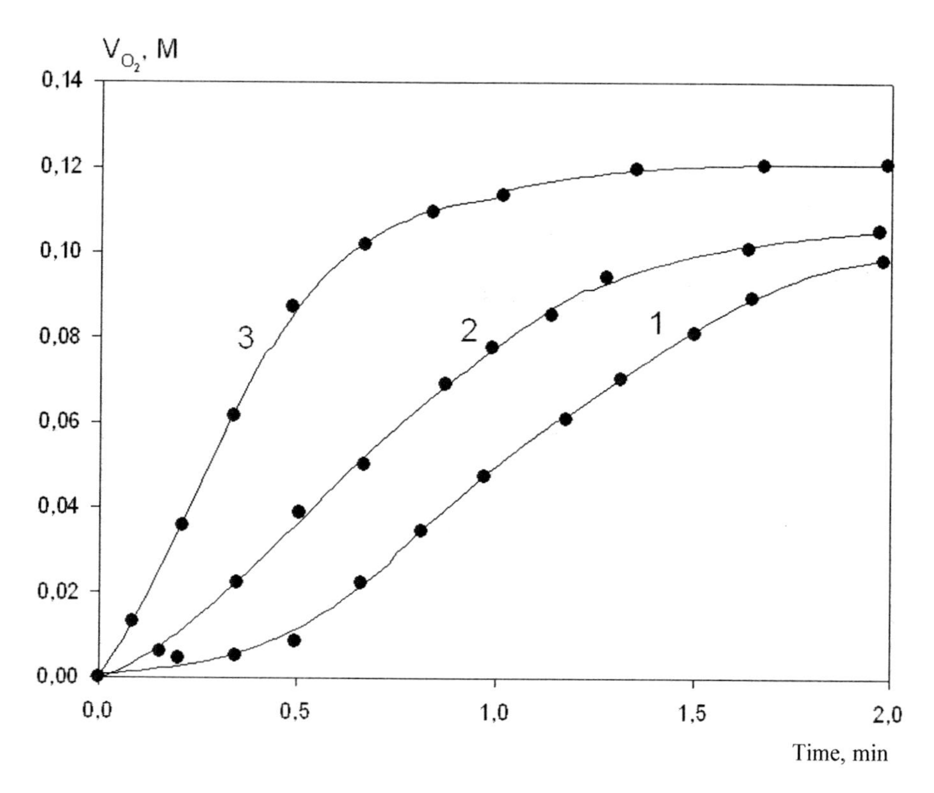

Figure 3. Kinetic curves of oxygen absorption in the course of process of conjugated oxidation of $SnCl_2{\cdot}2H_2O$ and cyclohexane: 1 – $[C_6H_{12}]$ – 0M, 2 – 0,02M, 3 – 0,5M. Conditions are: $[O_2]$=10 vol. %, $[H_2O]$=1,4M, $[Sn(II)]_0$=0,2M, temperature 25°C)

At stationary conditions at low water concentrations, when Sn (III) concentration is low, the main reaction of hydrogen peroxide and cyclohexyl consumption are reactions of interaction with bivalent tin chloride. In this case calculation gives the following expression for ratio of oxygen absorption rates in the presence and absence of cychexane:

$$\mathbf{W}^{\mathbf{RH}} / \mathbf{W}^0 = (1+\mathbf{BF})(1+\mathbf{F}), \tag{VII}$$

where

$$\mathbf{F} = \frac{1-\beta}{\beta'' + \dfrac{\mathbf{k}_{15}[\mathbf{Sn(II)}]}{\mathbf{k}_{16}[\mathbf{RH}]}} \cdot \frac{1+\dfrac{\mathbf{k}_{15}[\mathbf{Sn(II)}]}{\mathbf{k}_{16}[\mathbf{RH}]}}{1+\dfrac{\mathbf{k}_5[\mathbf{Sn(II)}]}{\mathbf{k}_{10}[\mathbf{RH}]}} \text{ and } \mathbf{B} = \frac{\alpha'\beta''}{\alpha\beta}.$$

The alcohol yield per initial concentration of divalent tin at that is:

$$\frac{[\mathbf{ROH}]_\infty}{[\mathbf{Sn(II)}]_0} = \frac{1-\beta}{1-\beta+2\beta''(1-\alpha')} - \frac{1-\beta}{\{1-\beta+2\beta''(1-\alpha')\}^2} \bullet \frac{\mathbf{k}_{15}[\mathbf{Sn(II)}]_0}{\mathbf{k}_{16}[\mathbf{RH}]} \tag{VIII}$$

We should note that formula (VIII) is approximate for low conversion degree of alkane, i.e. it doesn't take into account possibility of capture of a part of radicals $^{\bullet}OH$ and $^{\bullet}OR$ resulted from alcohol reaction. However, when comparing experimental results with kinetic curves calculated by formulas (VII) and (VIII) with the use of parameters $\alpha=5\cdot10^{-4}$, $\alpha'=3,6\cdot10^{-3}$, $\beta=0,85$, $\beta''=0,25$, $k_{15}/k_{16}=0,22$, $k_5/k_{10}=0,43$ we obtain good agreement.

Increase of alcohol yield with the rise of water concentration is caused by increase of consumption of Sn (III) reactions with hydroperoxides that leads to increase of efficiency of radicals $^{\bullet}OH$ ($^{\bullet}OR$) formation (reactions 4 and 14). With the rise of water concentration reactions of peroxides with Sn (II) (3) and (13) play less and less role. In water medium from $RO_2^{\bullet}$ ($HO_2^{\bullet}$) almost quantitatively $RO^{\bullet}$ ($HO^{\bullet}$) are formed. Expression (VIII) for this case is transformed into the following formula for calculation of relative alcohol yield:

$$\frac{[\mathbf{ROH}]_\infty}{[\mathbf{Sn(II)}]_0} = 1 - \frac{\mathbf{k}_{15}[\mathbf{Sn(II)}]_0}{\mathbf{k}_{16}[\mathbf{RH}]} \tag{IX}$$

At these conditions process becomes practically non-branched. Estimations show that alcohol is formed mainly as a result of long oxidation chain via radical $RO^{\bullet}$. Termination of this chain of hydrocarbon oxidation occurs as a result of radical $RO^{\bullet}$ capture by bivalent tin, alcohol or impurities. Length of chain may be very long:

$$\nu = \frac{\mathbf{k}_{16}[\mathbf{RH}]}{\mathbf{k}_{15}[\mathbf{Sn(II)}] + \mathbf{k}_{17}[\mathbf{ROH}]} \tag{X}$$

Maximum relative yield of alcohol per tin ($[ROH]_\infty/[Sn(II)]_0$) according with formula (IX) is equal to 1.

Analysis of suggested radical-chain mechanism of conjugated oxidation of cyclohexane and metals compounds [17] shows principle possibility of attainment of stoichiometry $[ROH]_\infty:[O_2]_\infty:[2e^-]$ = 1:1:1 characteristic for fermentative oxidation of alkanes with participation of monooxygenases (see section 3). This data indicate on the fact that stoichiometry of process can't be the proof of its mechanism.

In stoichiometric systems of cyclohexane oxidation conjugated with auto-oxidation of metals salts the yield of oxidation products per $[MCl_n]_0$ depends on metal nature (see Table 1). Relative yield of oxidation products may be increased by changing of structure of reaction mixture. So, the yields of products of cyclohexane oxidation per $SnCl_2$ – 20% and per $FeCl_2$ – 50% were experimentally obtained [11].

Usage of reducing agent transforming oxidized metal into initial state allows creating of alkanes oxidation' catalytic systems on the base of stoichiometric systems considered above [11]. Metals amalgams or free metals, and also electrons from cathode were used as reducing agents. In the case for catalytic systems the yield of cyclohexane oxidation products is significantly increased in comparison with stoichiometric systems. So, in catalytic system on the base of iron salt in the presence of reducing agent they were succeeded in receiving of 25 cycles of catalyst [11].

On the base of analysis of obtained results the important practical conclusions were made [17]. For creation of effective catalytic systems of selective radical-chain oxidation of alkanes at soft conditions it is necessary:

(1) to provide fast reaction proceeding via ion or metal complex to avoid slow stage $RO^{\bullet}_2 + RH$ and exclude quadratic termination of chains $RO^{\bullet}_2 + RO^{\bullet}_2$ with formation of set of products;
(2) to organize long chain of alkane oxidation via radical $RO^{\bullet}$ for providing of high yield of desired product $R\dot{O} \xrightarrow{RH} \dot{R} \xrightarrow{O_2} R\dot{O}_2 \xrightarrow{M} R\dot{O}$;
(3) to create catalytic conditions of reaction: to return ions or metals complexes into initial reduced state (for example electro-chemically).

Thus, on the base of studying of kinetics and mechanism of $SnCl_2$ and cyclohexane conjugated oxidation the basis of theory of low-temperature selective systems of radical-chain alkanes oxidation is developed. The possibility of attainment in radical-chain process of selectivity by product and stoichiometry typical for fermentative alkanes oxidation by monooxygenases is shown.

2. Catalytic Oxidation of Alkanes in Biomimetic Systems

Recently significant progress in creation of systems of catalytic oxidation of alkanes at soft conditions is appeared. Impressive results were obtained mainly due to development of relatively novel region of investigations – biomimetics [10, 18, 19]. The novel term "biomimetics" is proposed for old problem consisted in imitation of animate nature in creation of purely chemical (i.e. non-biological) systems which may realize various functions, for example, to catalyze various processes.

Advantages of animate nature chemistry over traditional chemistry were obvious for a lot of generations of scientists and caused attempts of their application (modelling). While imitating nature chemists develop methods of molecular design which also have their own advantages over animate nature – they are freer in selection of conditions and materials, more variable set of methods of influence on substance is at their service. Mastering of principles of living organisms functioning provides possibility of deliberated application of them in more simple molecularly-organized systems. This is the subject of biomimetic approach to solution of posed chemical problems. From the other hand, studying of biomimetic systems serve for extend one's knowledge about nature processes. Chemical systems while modelling only part of biological processes allows answering the questions about nature systems which are non-studying ones due to one or another reason.

2.1. Oxidation of Alkanes by Nature Monooxygenases

Ferments relating to class of monooxygenases catalyze alkanes oxidation by molecular oxygen in accordance with the equation:

$$RH + O_2 + 2e^- + 2H^+ \rightarrow ROH + H_2O.$$

Cytochrome P450 and methanemonooxygenase (MMO) are the most studied among these ferments [10]. Reaction cycle of these ferments is composed of two main processes: oxygen activation and hydrocarbon substrate oxidation. Spectral investigations allowed surely establishing of nature of intermediate complexes of cytochrome P450 catalytic cycle (Figure 4) beginning from initial Fe(III) to Fe(II)O_2 complex.

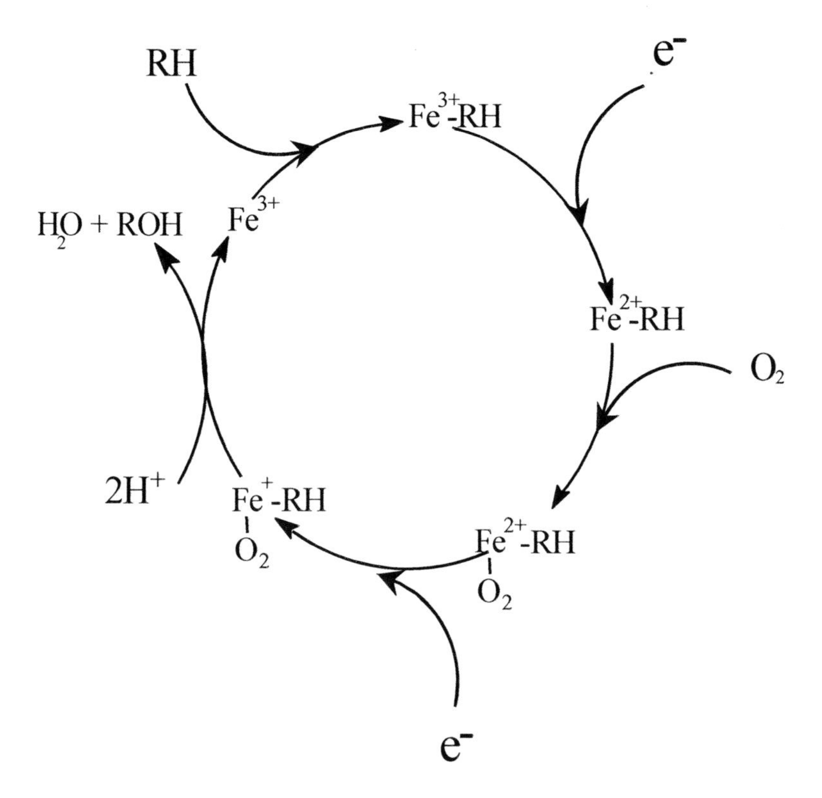

Figure 4. Catalytic cycle of Cytochrome P450.

Transfer of the second electron on oxycytochrome P450 leads to formation of $Fe^{2+}O_2^{-}$ or $Fe^{3+}O_2^{2-}$ particles. Intermediate stages following the reduction of oxycytochrome are too fast to detect intermediates by spectral methods. Mechanism of complex "substrate–reduced oxycytochrome" destruction is the least clear process in cytochrome P450 catalytic cycle. Usually they assume that formation of known oxidation products is caused by heterolytic or homolytic break of O–O bond in iron peroxo-complex. There are various variants of this mechanism and each of them is coordinated only with some part of accumulated experimental data. Basing on analogy with active intermediates of peroxidase catalytic cycle the majority of researchers prefer high-coval;ent complex of iron- porphyrin (formally PFe(V)=O or $P^{+\bullet}Fe(IV)=O$) as direct oxidizing agent for substrate. In spite of some differences catalytic cycle of MMO is analogue to catalytic cycle of cytochrome P450 in many respects. In catalytic cycle of MMO the structure of diferreel or bis-μ-oxocomplex is proposed for active particle.

In early works on studying of nature monooxygenases they proposed [20] direct entering of oxygen atom by C–H bond (so-called oxenoid mechanism). However later on for biological oxidation the crypto-radical mechanism got wide spread application [21], in other words the mechanism of removal-recombination also called in English-speaking literature the "oxygen rebound mechanism". According with this mechanism homolytic removal of H atom from hydrocarbon occurs with formation of radical $R^{\bullet}$ and hydroxyl group $^{\bullet}OH$ bonded with iron atom in active site. Molecule of alcohol is formed during following recombination reaction of $R^{\bullet}$ and $^{\bullet}OH$. One of the most cogent arguments in favour of this mechanism was the observed in many cases isomerization of substrate molecule significant enough which was explained by rotation of resulted radical. It seemed to be doubtless that direct entering of oxygen atom couldn't be accompanied by isomerization.

However, attentive consideration of experimental data on hydrocarbons oxidation by cytochrom P450 and its chemical models showed that crypto-radical mechanism didn't stand up to quantitative examination [22, 23]. The novel non-radical mechanism [22, 24] was proposed, its essential feature was formation of intermediate complex of substrate with active site. They assumed that this complex contained 5-coordinated carbon. Existence of two forms is possible for such complex:

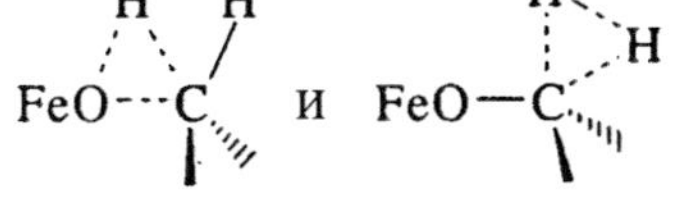

H H
FeO---C и FeO—C
H
H

The first of them may be considered as product of oxygen connection to bond C–H, for another one coming together of H atoms is characteristic and isomerization at the expense of H–H fragment turning is possible:

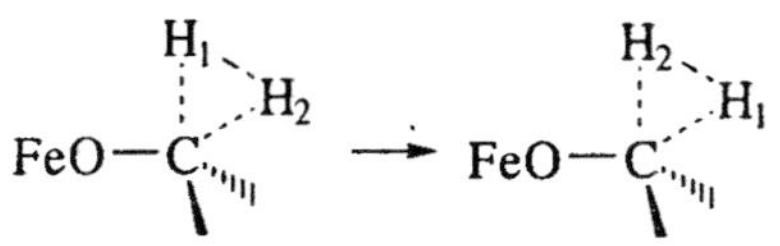

H_1 H_2
FeO—C → FeO—C
H_2 H_1

In the framework of suggested mechanism the data on isomerization of deuteron-substituted alkanes in the process of their hydroxylation with participation of monooxygenases were analyzed by the method of quasi-stationary concentrations [23-25]. Kinetic parameters of these reactions were calculated. Obtained results showed that non-radical mechanism of oxidation including connection of O atom by bond C–H with its further entering by this bond allowed with the help of quantitative characteristics explaining of unusual distribution of reaction products and isotope effects of hydroxylation of isotope-substituted molecules in fermentative systems, where together with hydroxylation the exchange of H and D atoms places occurred.

In spite of the large number of works devoted to discussion of mechanism of substrates hydroxylation by ferment system of methanotrophic bacteria – methanemonooxygenase (MMO) on the base of kinetic isotope effect (KIE) values researches took no notice of unusual result received under investigation of oxidation of soluble MMO of isotope marked methane (CH_3D, CH_2D_2, CHD_3) [26]. They found that values of KIE calculated by ratio of reaction products with consideration of a number of C–D and C–H bonds in $CH_{(4-n)}D_n$ molecule are significantly increased with the rise of deuterium atoms number (n). It is obvious from the framework of crypto-radical mechanism of methane oxidation by MMO that KIE = k_H/k_D and should not change with change of n.

Unusual fact of KIE values increase with the rise of deuterium atoms number was explained in the framework of non-radical mechanism (via formation of 5-coordinated carbon) [27]. Analysis of results of numerical experiment showed [28] that proposed kinetic models of methane oxidation were in good agreement with experimental data which didn't have explanation in framework of crypto-radical mechanism accepted by the majority of researchers.

Modeling of hemic monoxygenases significantly broaden possibilities of functionalization of non-activated C–H bonds promoting successful solution of the problem of rational usage of natural gas, oil and development of novel methods in thin organic synthesis. Such biomimetic approach allows also making conclusions about mechanisms of oxygen activation and oxidation of C–H bonds in fermentative and model monooxygenase systems. Results received with the use of biomimetic approach to creation of systems of hydrocarbons catalytic oxidation are presented below.

2.2. Chemical Models of Monooxygenases on the Base of Metals Porphyrinic Complexes

Since iron-porphyrinic complex enters the structure of cytochrome P450 active site it seems to be logically to use metal-porphyrins as catalysts under chemical modelling of monooxygenases. Attempts to model fermentative oxidation under the use of metal-porphyrin complexes and active oxygen donors such as iodosobenzene and its perfluorochemical analogous, sodium hypochlorite, hydrogen peroxide and alkylhydroperoxides, peracids, potassium persulfate, amines N-oxides, nitrogen protoxide turned to be very successful [10]. However, application of such effective oxidizing agents leads to contamination of reaction products by halogen-containing compounds and other impurities. From both ecological and economical points of view molecular oxygen is more preferable as oxidizing agent. Modelling of key process of monooxygenase catalysis – activation oligomerization of molecular oxygen on metals complex in the presence of electron donor (reducing agent) is of great practical and theoretical interests [29].

Synthesized from O_2 and PFe in the presence of reducing agent superoxo- and peroxoferril porphyrinic complexes are relatively stable compounds, inert in relation to alkanes. Only introduction into system of additional component – acetic acid – lead to creation of the first complete chemical model of cytochrome P450 on the base of PFe, active in relation to alkanes [30, 31]. Zinc powder or zinc amalgam was used in this system as reducing agent, as protons donor the acetic acid in acetonitrile solution was used. Alkanes are oxidized in this system at room temperature to corresponding alcohols, ketones and in some cases hydroperoxides [32, 33]. Introduction into this system of electron carrier methylviologen (MV) more than in order increases the rate of reaction of hydrocarbon oxidation [34]. Presence of substituents in phenyl cycles of ironporphyrin influences on both selectivity and rate of cyclohexane oxidation process in system with MV and without it. Selection of conditions allowed to receive 340 cycles of catalyst in 1 minute under oxidation of cyclohexane to alcohol and ketone in this system [34].

The main distinctions of proposed biomimetic systems from nature ones are in selectivity of reactions and catalysts stability. These characteristics for model catalytic processes depend on nature of oxidizing agent, substrate and conditions of reaction carrying out. Due to softness of conditions of reaction carrying out they were succeeded in avoiding of deep oxidation with formation of a large number of products. However, due to presence in alkanes of a large multitude of slightly differing between each other sites of attack high selectivity of their oxidation is hard-hitting.

Investigations showed [35], that two effects – electron and steric created by closest albuminous environment of active site in ferment molecule cause unusual selectivity in biological oxygenation. Alkanes molecules while connecting with albuminous matrix may have preferable orientations in relation to active site in such way that some C–H-bonds turned to be completely screened and other ones on the contrary have favorable for attack disposition.

One of the methods of modelling of catalyst micro-environment is its fixing on solid support, i.e. creation of heterogeneous oxygenase systems. Fixing of prophyrin complex of iron on porous glass [34] lead to significant increase of alcohol formation selectivity (more than 80%) under cycloalknes oxidation in the system on the base of molecular oxygen and donors of protons and electrons. As we expect, heterogenization of catalyst was accompanied by some reduce of rate of hydrocarbons oxidation. However developed heterogeneous oxygenase systems possess doubtless advantages over their homogeneous analogues. Essential disadvantage of the majority of known catalytic systems on the base of metal-porphyrins in solution is reduction of their activity in the course of oxidation reaction that limits the number of reacting cycles. Fixing of metal-porphyrins on various bearers was accompanied by both increase of selectivity of catalyzed by them oxidation processes and in the majority of cases by increase of catalysts stability. For example, fixing of iron-porphyrins on porous glass increases the number of reaction cycles of catalyst in system on the base of molecular oxygen almost in 200 times [34].

2.3. Active Intermediates in Metal-Porphyrinic Model Systems

Analogously to catalytic cycle of cytochrome P450 (Figure 4) formation of active intermediates in biomimetic systems on the base of molecular oxygen and electrons donor occurs at several consequent stages. The first stage is reduction of porphyrin. Reduced metal-porphyrins with high rate constant (about 10^7 $M^{-1}sec^{-1}$[36]) connect oxygen with formation of dioxygen complex. Presence of water in solution catalyzes destruction of dioxygen complexes with formation of radicals $HO_2^{\bullet}$. In water-less solvents complexes $PFe(II)O_2$ may be reduced to $(PFeO_2)^-$. For example, $TpivPPFe^{II}(O_2)(Tr\text{-}Im)$ is reduced in dehydrated acetonitrile solution by cation-radical of methylviologen (MV) with rate constant $k=(2{,}2\pm0{,}4)\cdot10^9$ $M^{-1}sec^{-1}$ [37].

Protonation of oxygen atom of peroxo-group may lead to homolytic or heterolytic break of bond O–O:

$$(PFeO_2)^- + 2H^+ \begin{cases} \xrightarrow{\delta} (PFeO)^+ + H_2O \\ \xrightarrow{1-\delta} (PFeOH)^+ + {}^{\bullet}OH \end{cases}$$

where δ and 1-δ are probabilities of heterolysis and homolysis correspondingly.

This reaction obviously is two-stage: the stage of the first proton connection is followed by the stage of protonation of intermediate compound iron-porphyrin hydroperoxide leading to break of O–O bond (analogously to systems on the base of hydrogen peroxide).

Kinetics and mechanism of active particles formation interacting with alkanes were studied in details in system metal-porphyrin / O_2 / AcOH / Zn. For answering on question about active particles number and their nature in this biomimetic system testing investigations were carried out [38]. Generally accepted tests stereoselectivity and kinetic isotope effect didn't give unambiguous answer on formulated question. Investigation of substrate selectivity under competitive oxidation of cyclopentane (S_5) and cyclohexane (S_6) turned to be more informative. Substrate selectivity was characterized by so-called parameter 5/6 calculated as ratio of corresponding constants of oxidation products formation rates with consideration of difference in a number of C–H bonds in substrates S_5 and S_6. It was established on the base of kinetic regularities that under oxidation of cycloalkanes in this system alcohols and ketones are formed parallel. That is why calculation of parameter 5/6 was carried out in three ways: by the ratio of initial rates of formation of alcohols $(5/6)_1$ and ketones $(5/6)_2$ and also by ratio of their sums $(5/6)_3$. Experiments showed that introduction into system of additional reagents (electron carrier, additional reducing agent, etc.) significantly changed the values of $(5/6)_1$ and $(5/6)_2$ whereas $(5/6)_3$ remained constant (in the limits of experiment error). The most probable candidates to direct reagents involving alkanes in oxidation process in this systems are radicals ${}^{\bullet}OH$ ($RO^{\bullet}$) and active iron-porphyrin complexes. If C–H bonds in cycloalkanes attack particles of one sort – A:

$$A + S_5 \xrightarrow{\alpha_5 k_{5S}} S_5OH;\ A + S_5 \xrightarrow{(1-\alpha_5)k_{5S}} S_5O;$$

$$A + S_6 \xrightarrow{\alpha_6 k_{6S}} S_6OH;\ A + S_6 \xrightarrow{(1-\alpha_6)k_{6S}} S_6O,$$

it's not hard to get the following expressions for substrate selectivity:

$$(5/6)_1 = 1{,}2\alpha_5 k_{5S}/(\alpha_6 k_{6S});\ (5/6)_2 = 1{,}2(1-\alpha_5)\,k_{5S}/[(1-\alpha_6)k_{6S}];\quad \text{(XI)}$$
$$(5/6)_3 = 1{,}2k_{5S}/k_{6S},$$

If alcohol and ketone are formed as a result of attack of C–H bonds by particles of different sorts – A_1 and A_2:

$$A_1 + S_5 \xrightarrow{k'_5} S_5OH;\ A_2 + S_5 \xrightarrow{k''_5} S_5O;$$
$$A_1 + S_6 \xrightarrow{k'_6} S_6OH;\ A_2 + S_6 \xrightarrow{k''_6} S_6O,$$

then:

$$(5/6)_1 = 1{,}2\ k'_5/k'_6;\ (5/6)_2 = 1{,}2\ k''_5/k''_6;\quad \text{(XII)}$$
$$(5/6)_3 = 1{,}2(k'_5[A_1] + k''_5[A_2])/(\ k'_6[A_1] + k''_6[A_2]),$$

where α_5 and α_6 – probabilities of formation of alcohols from S_5 and S_6; k_{5S}, k'_5, k''_5 and k_{6S}, k'_6, k''_6 – constants of rate of active particles A, A_1, A_2 interaction with substrates S_5 and S_6 accordingly.

Expressions (XI) and (XII) show that change of selectivity of alcohol and ketone formation in the first case (active particle A) should be accompanied by change of $(5/6)_1$ and $(5/6)_2$ at constant $(5/6)_3$. In the second case (active particles A_1 and A_2) parameter $(5/6)_3$should be changed under change of selectivity of process by products. Carried out analysis showed that further increase of number of active particles interacting with substrates led to change either of all three parameters of substrate selectivity with the change of alcohol and ketone formation selectivity, or only of parameter $(5/6)_3$ (in dependence on concrete mechanism of reaction).

Thus, experimentally observed constancy of parameter $(5/6)_3$ at significant change of $(5/6)_1$ and $(5/6)_2$ is the evidence in favor of mechanism of parallel formation of alcohol and ketone via the first common stage – reaction of interaction of some active particles A with substrate. Active particles possessing substrate selectivity differing from selectivity of particles A don't make significant contribution into products formation. At that true substrate selectivity (5/6) as it is obvious from expressions (XI) reflects the value of parameter $(5/6)_3$. Experimentally observed dependence of value of this parameter on nature of substituents in phenyl cycles of iron porphyrins and its value 0,38-0,50 characteristic for electrophilic attack by positively charged reagent [39] allow establishing that active particle in this system is iron-porphyrin complex [38]. Moreover, comparative investigation of selectivity of model substrate (anisole) oxidation carried out in systems of various types showed that in contrast to peroxo-complex of methylrhenium oxidizing anisole preliminary by C–H bonds of benzene cycle [40] iron-porphyrin system with molecular oxygen oxidized preliminary C–H bonds of methyl group of this alkylaromatic substrate. This fact testifies to less electrophilic character of active intermediate formed under reducing activation of molecular oxygen on iron-porphyrin site. On the base of analysis of obtained results the conclusion was made that active intermediates were high valency metal-pophyrine oxo-complexes.

Kinetic isotope effects and stereoselectivity of alkanes oxidation in biomimetic systems are strongly differed in dependence on nature of metal and substituents in phenyl cycles of metal-porphyrins, on nature of active oxygen donor and even on medium [10]. Comparison of regio-selectivity of microsomal oxidation of hexane in complete and shunted fermentative systems showed that nature of oxidizing agent (molecular oxygen and iodosobenzene) influenced on electrophilicity of active particles in nature systems [35]. Steric factor at that remains constant. For both nature and model systems the observed differences in test characteristics testify to some differences in structures of active intermediates obtained by various methods. It is obvious that degree of high valency oxo-complexes dissociation in polar and non-polar solutions may be different. Of course characteristics of such particles should have differences. Nucleophils or electrophils influence on structure of intermediates in analogous way.

2.4. Mechanism of Alkanes Oxidation in Biomimetic Systems

Mechanism of oxygen atom transfer by porphyrin complexes on C–H bonds in biomimetic systems as well as in nature ones is the subject of discussions. Differences in structure of active metal-porphyrin complexes obtained by various methods lead to differences in mechanism of alkanes oxidation by these complexes. At that isomerization of oxidation substrate in both biomimetic and nature systems is not the proof of intermediate formation of alkyl radicals.

For example, hydroxylation of (R)- or (S)-enantiomers of monodeuteroethylbenzene by donor of active oxygen (iodosobenzene) under catalysis by chiral porphyrin iron complex leads to formation of all possible isomers of 1-phenylethanol [41]. At that the main product in both cases is formed stereoselectively and represents practically pure monodeuterophenylethanol, where as the product formed in fewer amounts is only half-deuterized. This experimental fact couldn't be explained in the framework of crypto-radical mechanism of removal-recombination. At the same time complete agreement between observed and calculated in the framework of crypto-radical mechanism (see kinetic scheme of hydroxylation of (R)-monodeuteroethylbenzene via formation of 5-coordinated carbon in Figure 5) values for relative yields of products of hydroxylation of enantiomers of partially deuterated ethylbenzene was reached (Table 2) [25]. Oxidation of chiral ethylbenzene is the example for which change of value of rate constant of hydroxylation (in given case at the expense of high KIE) leads to practically complete saving of configuration (for C–H bond) and simultaneously to significant isomerization (for C–D bond) in one and the same molecule. This example helps to understand in what way complete saving of configuration or its conversion (i.e. racemization) may be observed in very similar systems. With the help of kinetic parameters calculated for example by experimental data for R-isomer of deuteroethylbenzene without any additional postulates we may accurately predict values obtained experimentally for S-isomer [23]. This fact is convincing argument in favor of proposed mechanism via intermediate formation of 5-coordinated carbon.

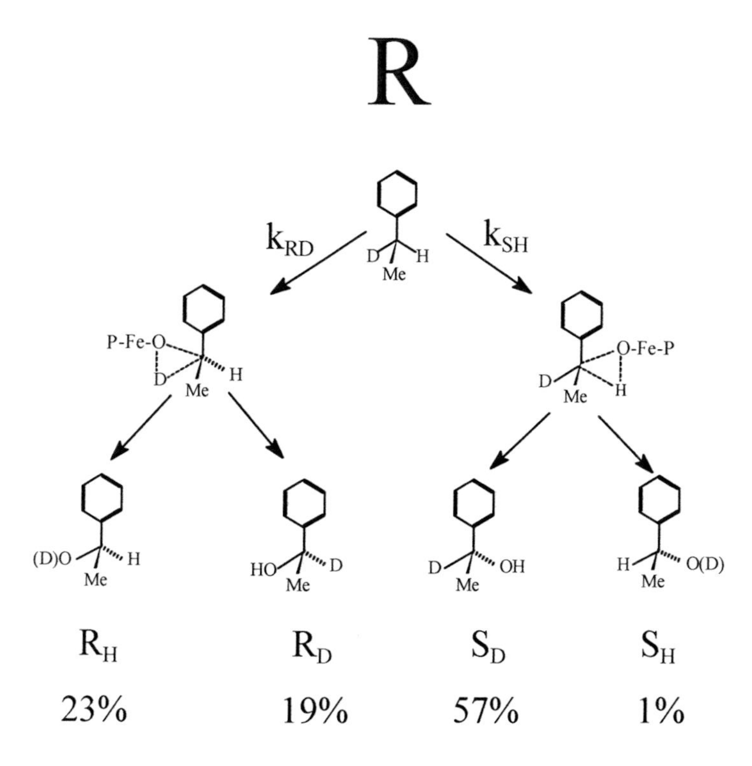

Figure 5. The scheme of hydroxylation of (R)-monodeuteroethylbenzene via formation of 5-coordinated carbon.

Table 2. Relative yield of isomers of 1-phenylethanol and its deuteroisotopomer under oxidation of (R)- and (S)-(1-deuteroethyl)benzene in biomimetic system with donor of active oxygen – iodosobenzene.

Substrate	Isomer*	Relative yield, %	
		Experiment [41]	Calculation [23]
(R)-(1-deuteroethyl)benzene	R_H	23	22,9
	R_D	19	19,1
	S_H	1	1,1
	S_D	57	56,9
(S)-(1-deuteroethyl)benzene	R_H	2	1,7
	R_D	87	87,3
	S_H	6	6,0
	S_D	5	5,0

* Indexes H and D are related to products of C–H and C–D bonds oxidation accordingly.

As we have already mentioned biomimetic systems on the base of molecular oxygen were usually less selective than systems with donors of active oxygen such as iodosobenzene and sodium hypochlorite. For example in system $FeP/O_2/Zn/AcOH/CH_3CN$ from cyclohexanone together with alcohol the ketone and small amount of hydroperoxide are formed (Figure 6) [32, 33]. Formation of noticeable amounts of cyclohexyl hydroperoxide indicates on the fact that in this system alkanes oxidation even partially occurs by radical route.

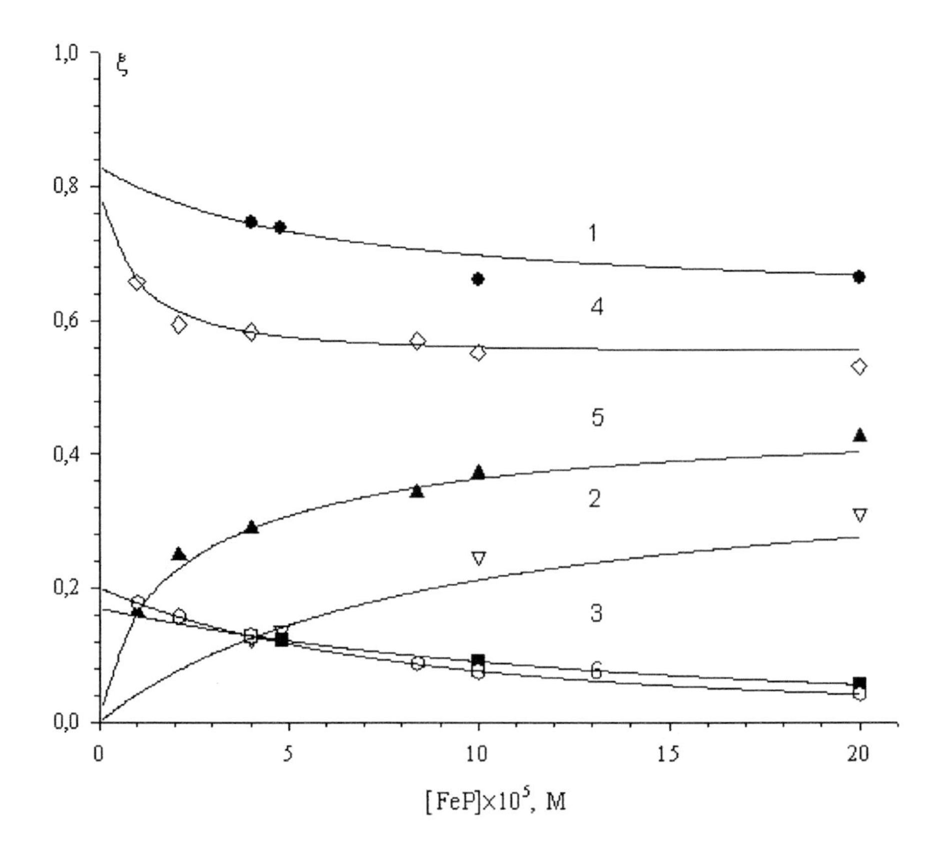

Figure 6. Dependence of parameters of selectivity $\xi=[P]/\Sigma[P]$ (1,4 – $\xi=x=[ROH]/\Sigma[P]$; 2,5 – $\xi = y = [R'O]/\Sigma[P]$; 3,6 – $\xi = z= [ROOH]/\Sigma[P]$) on $[FeTDCPP]_0$ under oxidation of C_6H_{12} by molecular oxygen in biomimetic system. Conditions of reaction are: temperature 20°C; 0,1 MPa of air; CH_3CN; 30 mg Zn; $[C_6H_{12}]$ =0,7M; $[CH_3COOH]$ =0,1M; curves 1-3: $[MV]=7{,}5\cdot10^{-3}M$; curves 4-6: [MV]=0).

Kinetic scheme of oxygen activation on iron-porphyrin in biomimetic system $FeP/O_2/Zn/AcOH/CH_3CN$ [34] represents consecution of oxidation-reduction reactions of catalytic cycle analogously to cycle of cytochrome P450. Active particles in this system are compl,exes of iron porphyrins [38]. These complexes $(FeO)^+$ may interact with saturated hydrocarbons with formation of radical or molecular products:

$$18.\quad RH + (FeO)^+ \longrightarrow \begin{cases} \xrightarrow{\alpha_1} R^\bullet + {}^\bullet OH + (\overset{III}{Fe}{}^+) \\ \xrightarrow{1-\alpha_1} (Fe^{IV}\text{-}\underset{+}{\overset{H}{O}}\text{-}R) \end{cases}$$

Positively charged complex of active particles with hydrocarbon may either decompose with formation of alcohol, or interact with some reagents and form ketone or alcohol. For simplicity of consideration we assume that probability β_1 of alcohol formation by molecular route takes into account these both possibilities:

$$19.(Fe^{IV}\text{-}\overset{\overset{H}{|}}{\underset{+}{O}}\text{-}R) \longrightarrow \begin{cases} \xrightarrow{\beta_1} ROH \\ \xrightarrow[1-\beta_1]{} R'O \end{cases}$$

Release of $R^{\bullet}$ and $^{\bullet}OH$ radicals into volume at aerobic conditions in the presence of donors of protons and electrons will lead to formation of hydroperoxide:

$$RH + {}^{\bullet}OH \to R^{\bullet}$$

$$20'.\ R^{\bullet} \xrightarrow{O_2} RO_2^{\bullet} \xrightarrow{e^-,H^+} ROOH$$

Hydroperoxide may destructed in reaction mixture to alcohol and ketone under interaction with various intermediate complexes of iron-porphyrin:

$$21.\ ROOH + (A) \longrightarrow \begin{cases} \xrightarrow{\gamma} ROH + \text{продукты} \\ \xrightarrow{1-\gamma} R'O + \text{продукты} \end{cases}$$

where (A) – ironporphyrin complexes.

Presence of protons and electrons donors in reaction mixture assumed also proceeding of the following reaction:

$$22.\ {}^{\bullet}OH \xrightarrow{e^-,H^+} H_2O$$

Calculation of kinetic scheme of hydrocarbon oxidation in quasi-stationary approximation gives the following expressions for relative yields of products of radical route of cyclohexane oxidation (hydroperoxide and resulted from its decomposition alcohol and ketone – accordingly z, x_r and y_r) at the time moment t:

$$\begin{cases} x_r = \dfrac{d[ROH]_r}{d\sum\limits_{i=1}^{5}[P_i]} = \alpha_1''\gamma\cdot\left\{1+\dfrac{\exp(-k_{21}^{ef}[FeP]_0t)-1}{k_{21}^{ef}[FeP]_0t}\right\} \\ y_r = \dfrac{d[R'O]_r}{d\sum\limits_{i=1}^{5}[P_i]} = \alpha_1''(1-\gamma)\cdot\left\{1+\dfrac{\exp(-k_{21}^{ef}[FeP]_0t)-1}{k_{21}^{ef}[FeP]_0t}\right\} \\ z = \dfrac{d[ROOH]_r}{d\sum\limits_{i=1}^{5}[P_i]} = \dfrac{\alpha_1''\{1-\exp(-k_{21}^{ef}[FeP]_0t)\}}{k_{21}^{ef}[FeP]_0t} \end{cases} \quad \text{(XIII)}$$

where $\mathbf{k}_{21}^{ef} = \mathbf{k}_{21}[(A)]/[FeP]_0$, $\alpha_1'' = \dfrac{\alpha_1'}{1-\alpha_1+\alpha_1'}$

$$\alpha' = \frac{2k_{20}[RH]+k_{22}}{k_{20}[RH]+k_{22}} \times \alpha_1$$

It is consequent from definitions of α'' and α' that at $k_{22} >> 2k_{20}[RH]$ the value is $\alpha''=\alpha'=\alpha$.

In accordance with equation (XIII) relative part of oxidation products formed by radical route ($P_{r,i}$) is determined by the expression:

$$\frac{\sum_{i=1}^{3}[P_{r,i}]}{\sum_{i=1}^{5}[P_i]} = z + x_r + y_r = z \cdot \frac{k_{21}^{ef}[FeP]_0 t}{1-\exp(-k_{21}^{ef}[FeP]_0 t)} = \alpha_1'' \tag{XIV}$$

Thus, relative part of hydrocarbon oxidation products resulted by radical way is equal to value α''_1. As we have already mentioned at $k_{22} >> 2k_{20}[RH]$ the value is $\alpha''_1=\alpha'_1=\alpha_1$, i.e. the part of oxidation products resulted via radical route is equal to probability of release of alkyl radicals into volume by reaction (18). If this inequality is not applied the value α''_1 is higher than value α_1, since in this case reaction of radicals $^{\bullet}OH$ interaction with hydrocarbon makes contribution into hydroperoxide formation process.

In accordance with expressions (XIII) and taking into account experimental values of effective constant of rate of hydroperoxide decomposition ($\mathbf{k}_{21}^{ef}$) [32] the value α''_1 may be calculated by values of current concentration of hydroperoxide in relative units (z) corresponding to moment of time t and initial concentration of iron-porphyrin $[FeP]_0$.

Analysis of experimental data showed [32] that for investigated iron-porphyrins the part of oxidation products resulted by radical route didn't exceed 20% from total yield of products. In this biomimetic system the value is $\alpha''_1 \approx \alpha_1$, i.e. interaction of hydroxyl radicals with hydrocarbon doesn't make essential contribution into reaction kinetics that seems to be natural in the presence of significant amounts of electrons and protons in reaction mixture. Radical route is realized at the expense of release of alkyl radicals into volume as a result of reaction 18 and forms small part of cyclohexane oxidation process.

Thus, both alcohol and ketone in system $FeP/O_2/Zn/AcOH/CH_3CN$ are mainly formed by molecular route. Alcohol at that may be formed directly by reaction of decomposition of active intermediate complex with hydrocarbon. At the same time for formation of ketone this complex should interact with other components of reaction mixture [33] that intend the lifetime of this complex is long enough.

Analysis of investigations results of alkanes oxidation in biomimetic and nature systems allows formulating of some general principles of creation of organized molecular systems – models of monooxygenases. Optimal rate and selectivity in biomimetic systems may be obtained if the following postulates are realized:

(1) realization of concrete monomolecular mechanism of oxidation of substrate instead of usual biomolecular. With the help of catalyst, placing functional groups of oxidizing agent and substrate in immediate proximity to each other one can significantly increase selectivity and rate of their interaction. This principle of spatial proximity is common for fermentative catalysis;
(2) process rate should be limited by stage of interaction of substrate with active particle. Preliminary stages of reducing activation of oxidizing agent may be accelerated either by fixing of catalytic complex close to electron donor, or by introduction of electron carriers into system. The last ones in optimal case should be spatially placed between electron donor and catalyst;
(3) it is desirable to realize with the help of catalyst the transition of one-electron process provided by one-electron donor to the two- or more electron reduction (oxidation) of substrate by catalyst. At that process thermodynamics allows using of weaker reducing agent that reduces energy consumption;
(4) oxidation of substrate should be realized conjugatevly with oxidation of external donor. Energy of oxidation of this donor may be used for creation of especially active intermediate catalyst forms able to oxidize even comparatively inert molecules of substrate.

In organized molecular systems – models of oxygenases the mechanism of external donor action consists in reduction of catalyst particle, and then oxidation of this particle for example by oxygen creates active catalytic site. At that not only conjugation with oxidation of donor, but also the conjugation with acid-base process also providing creation of active particle may be realized. For overcoming of unproductive reactions in such process (for example reaction of electron donor with oxidizing agent or acid-base interaction not conjugated with oxidation-reduction reaction) the organized ensemble may be used, for example by placing of catalyst between donor and acceptor.

The presented review clearly testifies to perspectives of Professors N.N. Semenov's and N.M. Emanuel's ideas application for developing of ecologically pure and resource-saving technologies. Systems of soft oxidation of saturated hydrocarbons considered in review demonstrate principle possibility of creation of high selective process of alkanes functionalization. Fundamental knowledge of processes mechanism allows later purposeful creating of more perfect systems of selective obtaining of valuable hydrocarbon products. Presented results of investigations of chemical modelling of hemic monooxygenases earnestly show that usage of principles of animate nature for creation of organized chemical systems is perspective approach to creation of technologies of future.

REFERENCES

[1] N.M. Emanuel, E.T. Denisov, Z.K. Maizus, Chain reactions of hydrocarbons oxidation in liquid phase, Moscow: *Nauka* (1965) (in Russian).
[2] N.M. Emanuel, G.E. Zaikov, Z.K. Maizus, The role of medium in radical-chain reactions of organic compounds oxidation, Moscow: *Nauka* (1973) (in Russian).
[3] A.N. Bakh, ZhRFKhO, 29, 373 (1897).

[4] N.N. Semenov, Chain reactions, Leningrad: *Goskhimtekhizdat* (1934) (in Russian).
[5] N.N. Semenov, *On some problems of chemical kinetics and reaction ability*, Moscow: Izd-vo AN SSSR (1958) (in Russian).
[6] N.M. Emanuel, *Uspekhi khimii*, 47, No.8, 1329, (1978) (in Russian).
[7] N.M. Emanuel, Izv. AN SSSR, *Ser. Khim.*, No.5, 1056 (1974) (in Russian).
[8] E.T. Denisov, N.M. Emanuel, *Uspekhi khimii*, 29, No.12, 1409 (1960) (in Russian).
[9] I.V. Berezin, E.T. Denisov, N.M. Emanuel, *Cyclohexane oxidation*, Moscow: Izd-vo MGU (1962) (in Russian).
[10] E.I. Karasevich, V.S. Kulikova, A.E. Shilov, A.A. Shteinman, *Uspekhi khimii*, 67, No.4, 376 (1998) (in Russian).
[11] E.I. Karasevich, N.Z. Muradov, A.A. Shteinman, Izv. AN SSSR, *Ser. Khim.*, No.8, 1805 (1974) (in Russian).
[12] I.V. Zakharov, E.I. Karasevich, A.E. Shilov, A.A. Shteinman, *Kinetika i kataliz*, 16, No.5, 1151 (1975) (in Russian).
[13] Yu.V. Geletii, I.V. Zakharov, E.I. Karasevich, A.E. Shilov, A.A. Shteinman, In coll.: Oxidation-reduction reactions of free radicals, Kiev: *naukova dumka* (1976).
[14] Yu.V. Geletii, I.V. Zakharov, E.I. Karasevich, A.A. Shteinman, *Kinetika i kataliz*, 20, No.5, 1124 (1979) (in Russian).
[15] E.A. Kutner, B.P. Matzeevskii, *Kinetika i kataliz*, 10, No.5, 997 (1969) (in Russian).
[16] V.F. Shuvalov, A.P. Moravskii, Doklady AN SSSR, 234, No.6, 1402 (1977) (in Russian).
[17] Yu.V. Geletii, E.I. Karasevich, A.A. Shteinman, *Neftekhimiya,* 18, No.5, 673 (1978) (in Russian).
[18] E.I. Karasevich, A.E. Shilov, *Khimiya v interesakh ustoichivogo razvitiya*, 8, No.4, 515 (2000) (in Russian).
[19] E.I. Karasevich, *Ross. Khim. Zh.*, 39, No.1, 31 (1995) (in Russian).
[20] G.A. Hamilton, *J. Am. Chem. Soc.*, 86, No.16, 3391 (1964).
[21] J.T. Groves, *J. Chem. Educ.,* 62, No.I, 928 (1985).
[22] A.F. Shestakov, A.E. Shilov, *J. MoL Catal.*, 105, No.I, 1 (1996).
[23] E.I. Karasevich, A.F. Shestakov, A.E. Shilov, *Kinetika i kataliz*, 38, No.6, 852 (1997) (in Russian).
[24] E.I. Karasevich, A.E. Shilov, *Exp. Toxic. Pathol.*, 50, No.2, 112 (1998).
[25] E.I. Karasevich, A.F. Shestakov, A.E. Shilov, *Exp. Toxic. Pathol.*, 51, No.4-5, 335 (1999).
[26] J.C. Nesheim, J.D. Lipscomb, *Biochemistry,* 35, No.31, 10240 (1996).
[27] E.I. Karasevich, Yu.K. Karasevich, A.F. Shestakov, A.E. Shilov, *Kinetika i kataliz*, 44, No.1, 122 (2003) (in Russian).
[28] E.I. Karasevich, Yu.K. Karasevich, A.F. Shestakov, A.E. Shilov, *Kinetika i kataliz*, 44, No.2, 266 (2003) (in Russian).
[29] A.E. Shilov, E.I. Karasevich, Oxidation of alkanes, In: *Metalloporphyrins catalyzed oxidations.*, Ed. Montanary F., Casella L. Netherlands: Kluwer Academic Publishers (1994).
[30] E.I. Karasevich, A.M. Khenkin, A.E. Shilov, Doklady AN SSSR, 295, No.3, 639 (1987) (in Russian).
[31] E.I. Karasevich, A.M. Khenkin, A.E. Shilov, *Chem. Soc., Chem. Comm.*, No.10, 731 (1987).

[32] E.I. Karasevich, Yu.K. Karasevich, *Kinetika i kataliz*, 41, No.4, 535 (2000) (in Russian).
[33] E.I. Karasevich, Yu.K. Karasevich, *Kinetika i kataliz*, 41, No.4, 543 (2000) (in Russian).
[34] E.I. Karasevich, B.L. Anisimova, V.L. Rubailo, A.E. Shilov, *Kinetika i kataliz*, 34, No.4, 650 (1993) (in Russian).
[35] E.I. Karasevich, A.M. Khenkin, *Bioklhimiya*, 51, No.9, 1454 (1986) (in Russian).
[36] T.G. Traylor, P.S. Traylor, *Annu. Rev. Biophys. Bioenerg.*, No.11, 103 (1982).
[37] A.B. Belyaev, E.I. Karasevich, V.A. Kuz'min, P.P. Levin, A.M. Khenkin, *Izv. AN SSSR, Ser. Khim.*, No.8, 1877 (1987) (in Russian).
[38] E.I. Karasevich, B.L. Anisimova, V.L. Rubailo, A.E. Shilov, Kinetika i kataliz, 34, No.4, 656 (1993) (in Russian).
[39] E.S. Rudakov, reactions of alkanes with oxidizing agents, metal-complexes and radiclas in solutions, Kiev: *Nauk. dumka* (1985).
[40] E.I. Karasevich, A.V. Nikitin, V.L. Rubailo, *Kinetika i kataliz*, 35, No.6, 878 (1994) (in Russian).
[41] J.T. Groves, P. Viski, *J. Am. Chem. Soc.*, 111, No.12, 8537 (1989).

In: Reactions and Properties of Monomers and Polymers
Editors: A. D'Amore and G. Zaikov, pp. 69-90
ISBN: 1-60021-415-0

Chapter 4

PHOTO-INDUCED BONDING OF WATER AND ALCOHOLS TO 1,2-DIHYDROQUINOLINES. MECHANISM OF REACTION AND PARTICULARITIES OF INTERACTION OF FORMED CARBOCATIONS WITH ION AND MOLECULAR NUCLEOPHILS

T. D. Nekipelova
N.M. Emanuel's Institute of Biochemical Physics Russian Academy of Sciences
4, Kosygin str., Moscow 111999, Russia

INTRODUCTION

Professor Nikolai M. Emanuel was organizer and permanent leader of laboratory of oxidation of organic compounds in liquid phase. Works on this theme represented the novel grade in scientific life of the Institute of Chemical Physics of USSR Academy of Sciences, since the main investigations in the Institute were connected with gas-phase and solid-phase reactions. Group of young scientists in majority recent graduates of Department of chemical kinetics of MSU was singled out in the structure of this laboratory in sixties of the past century the leadership of which was committed to N.M. Emanuel's follower A.B. Gagarina.

The broad research area directed on revealing of kinetics and mechanism of various classes compounds oxidation in connection with particularities of their structure was suggested for this group and in all investigations to a significant extent the antioxidants were used as a mean for studying of elementary stages of chain process in particular for determination of initiation rate. Independent problem was broadening of anti-oxidants range, revealing of detailed mechanism of their action in connection with practical demands of stabilization of easy oxidized substances.

In particular, the mechanism of action of anti-oxidant of hydrogenated quinoline class – 6-ethoxy-2,2,4-trimethyl-1,2-dihydro quinoline (ethoxychin) – was studied in details in a wide temperature interval [1-7]. They showed that particularity of this anti-oxidant highly effective at moderate temperatures was in the fact that at temperatures higher than 100°C ethoxychin was oxidized by molecular oxygen by chain radical mechanism including destruction of aminil radical to corresponding quinoline and methyl radical as limiting stage of chain propagation, and ethoxychin interaction with resulted methyl hydroperoxide was degenerated branching of chain [1-4]. Investigation of mechanism of this process allowed explaining of regularities of ethoxychin inhibitory action in broad temperature interval in various oxidizing substrates. Received mechanism of ethoxychin radical transformations allowed forecasting of their inhibiting action at various conditions [5-7] and suggesting the ways of modifying of ethoxychin for improvement of its activity at high temperatures.

Interest to hydrogenated quinolines was caused first of all by the fact that the products of acetone condensation with aniline and its derivatives 2,2,4-trimethyl-1,2-dihydroquinolines (DHQ) with various substituents in position 6 [8, 9] were the effective inhibitors of chain oxidation of organic compounds especially unsaturated ones [10]. These products under various brands were widely used long ago in many World countries as effective into-oxidants and anti-ozonants for rubbers, propellants and lubricating oils [11-13]. These compounds are practically non-toxic [14-17], that is why they are used for stabilization of β-carotin, vitamin A and other poly-unsaturated substances in fodders, for processing of vegetables and fruits before storage [18]. A lot of dihydroquinolines are biologically active substances and are studied as medical preparations [6, 17, 19-23].

We should note structural similarity of dihydroquinolines with indoles which enter as fragments into structure of a lot of biologically important compounds, for example amino-acids of tryptophan. Moreover, dihydroquinolines with various substituents in aromatic cycle and heterocycle turned to be suitable objects for studying of elementary radical reactions and photo-chemical transformations of nitrogen-containing heterocycles including such important reactions as proton transfer in exited and basic state. A lot of dihydroquinolines are biologically active substances, so results of investigations of reactions in proton solvents especially in water and micellar solutions are of great interest.

Alkyl and alkoxy-1,2-dihydroquinolines are the examples of compounds for which direction, mechanism and kinetic regularities of photolysis developing are extremely sensitive to medium's nature at all stages of reaction, from primary photo-physical and photo-chemical processes to thermal reactions of key intermediate products, aminil radicals and carbcations. Kinetic approach to solution of problems of reactions mechanisms studying developed under direction of Professor N.M. Emanuel was used under investigation of detailed mechanism of reactions of hydrogenated quinolines under the action of light. Results of this work made in the Laboratory of photosensitization processes of N.M. Emanuel's Institute of biochemical physics RAS are presented in given Chapter. The main attention is paid to establishing of mechanism of dihydroquinolines photolysis in proton solvents and reactions of carbcations generated under their photolysis in these solvents.

1. Experimental Part

The following alkyl- and alkoxy-substituted 2,2,4-trimethyl-1,2-dihydro quinolines were used I this work:

- secondary DHQ: 2,2,4-trimethyl-1,2-dihydro quinoline (acetonanil, 6-H-DHQ);

- 6-methyl-2,2,4-trimethyl-1,2-dihydro quinoline (6-Me-DHQ);
- 6-ethoxy-2,2,4-trimethyl-1,2-dihydro quinoline (ethoxychin, 6-EtO-DHQ);
- 8-methoxy-2,2,4-trimethyl-1,2-dihydro quinoline (8-MeO-DHQ);
- 6-nitro-2,2,4-trimethyl-1,2-dihydro quinoline (8-NO_2-DHQ);
- tertiary DHQ: 1,2,2,4,6-pentamethyl-1,2-dihydro quinoline (1,6-diMe-DHQ).

Methods of DHQ synthesis and investigation are described in [24-27] in details. For registration of absorption spectra and kinetics of destruction of short-term living intermediate products we used the methods of usual and laser impulse photolysis with spectrophotometric registration. Reaction was carried out in regime of stationary photolysis for isolation and analysis of products. Reaction course was controlled by UV spectra. Isolated products of DHQ photolysis were analyzed by IR-, UV-, ^{1}H and ^{13}C NMR spectroscopies and mass-spectroscopy. All rates constants are determined with 10% error and are presented in the text for temperature equal to 20°C (if another temperature is not given).

2. Mechanism of Photo-Induced Transformations of Dihydroquinolines

2.1. Photo-Transformations of Dihydroquinolines in Hydrocarbon Solutions

Initially investigation of DHQ transformations under the action of light was undertaken with the aim of determination of rate constant of recombination of aminil radicals resulted from DHQ oxidation, since it was known that secondary aromatic amines to which the dihydroquinolines without substituents at nitrogen atom might be ascribed to, under photolysis in organic solvents underwent homolytic break of N–H bond and formed aminil radicals analogous to those that resulted from dark radicals reactions with participation of these compounds [28]. Actually, we showed that under photo-excitation of 1,2-dihydroquinolines in hydrocarbon solutions and alcohols with $C \geq 2$ under the action of light the aminil radicals $RN^{\bullet}$ analogous to those ones resulted from DHQ oxidation were formed [29]. Kinetics of destruction of these intermediate substances obeys the law of the second order. As a result of photolysis and under oxidation of 6-EtO-DHQ the 1-8 dimer is formed [30, 31]. Biomolecular rate constants of destruction of $RN^{\bullet}$ from ethoxychin (6-EtO-DHQ) and acetonanil (6-H-DHQ) in hexane and benzene are equal to the values higher than 3×10^7 mole $l^{-1}sec^{-1}$ [29]. High constant of recombination rate of aminil radicals from DHQ leads to the fact that stoichiometric coefficient of inhibition by dihydroquinolines only insignificantly exceeds 1 [5, 32].

It is known from literature [10] that in contrast to other effective anti-oxidants the ethoxychin keeps high activity under oxidation of linoleic acid in water micellar solutions of sodium dodecylsulfate. Stoichiomteric coefficient of inhibition f is increased in micellar solutions up to 2. Such rise of f may be connected with reduction of constants of recombination rate of aminil radicals formed from ethoxychin in water solutions. So, the attempt was made to determine this rate constant in water and micellar solutions by method of impulse photolysis. However, we were not succeeded in this since it turned out that under direct photolysis of DHQ in water direction of reaction was sharply changed.

2.2. Products of Photolysis of Dihydroquinolines in Water and Methanol

Investigation of DHQ photolysis in water and methanol showed that these solvents served as reagents and were added by double bond of DHQ in accordance with Markovnikov's rule with formation of 4-hydroxy- or 4-methoxy-2,2,4-trimethyl-1,2,3,4-tetrahydroquinolines (Scheme 1). Reaction proceeds with yields close to quantitative [24-26]. In contrast to MeOH in water-less EtOH and Pr^nOH photo-reaction proceeds the same way as in other organic solvents with formation of dimer. Addition of water or methanol leads to transition from formation of aminil radicals recombination products to reaction of formation of mixture of alkoxy-and hydroxy- or two alkoxy-adducts. Thus, ethoxy- and propyloxy-adducts may be formed only in mixture with hydroxyl- or methoxy-adducts. Isopropanol doesn't form adducts due to steric hindrances.

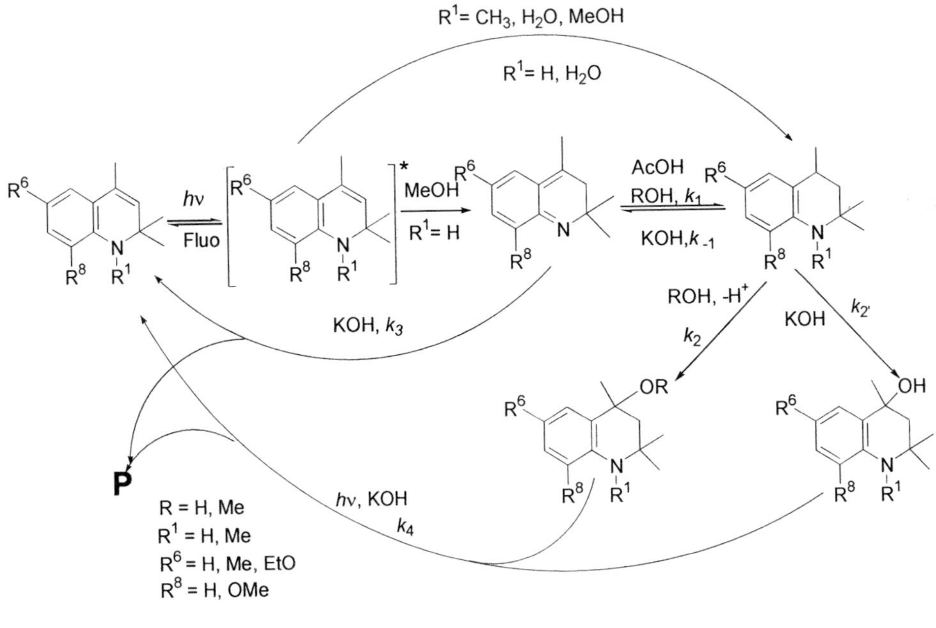

Scheme 1.

The products were isolated and identified on the base of UV, IR, ^1H and ^{13}C NMR spectra and mass-spectrometry. Physical-chemical parameters and spectra of 17 synthesized for the first time tetrahydroquinolines with hydroxy-, methoxy-, ethoxy- and propyloxy-substituents in position 4, and also adduct with azide ion are presented in [24-27]. Comparison of products of DHQ photolysis in methanol and products of initiated thermal oxidation of DHQ in methanol unambiguously showed that under photolysis in MeOH formation of aminil radical didn't occur [24].

2.3. Spectral-Kinetic Regularities and Mechanism of 1,2-Dihydroquinolines Photolysis in Methanol and Water

Kinetic regularities of photo-induced addition of water and methanol were investigated by the method of impulse photolysis in wide diapason of concentrations of added acid and alkali at various temperatures [26, 33-35]. Under photo-excitation of long-wave absorption band of DHQ (λ_{max} is 360-330 nm in dependence on DHQ and solvent) the process is one-quantum and occurs from excited state S_1. Under photolysis of secondary DHQ in MeOH two intermediates with λ_{max} 420 and 480-500nm correspondingly are formed consecutively, and addition of acid leads to faster formation of the second intermediate, and strongly alkaline solutions the formation of only first intermediate is observed. Adequate approximation of experimental kinetic curves of decline of intermediate optical density at various waves lengths was obtained while assuming that reaction proceeded in two consecutive stages (A $\rightarrow$ B $\rightarrow$ C, where A – was the first intermediate particle, B – the second intermediate particle, and C – final product). As a result of simultaneous processing of the whole massif of kinetic curves (global kinetic analysis) obtained under registration at waves lengths from 400 up to 600nm in accordance with this scheme the values of rates of the first and second stages k_1 and k_2 presented in [34] were calculated. Life times of intermediate particles are increased with the rise of electron-donor ability of substituents in aromatic nucleus. Maximum life-time of particle A in neutral solution of MeOH was observed for 6-EtO-DHQ (40 msec) and minimum for 6-H-DHQ (0,7 msec). For compounds with substituents in position 6 the activation energy of the first reaction is close to 0, whereas in the case of 8-MeO-DHQ it is $E_1^a = 15{,}0$ kJ/mole.

In the case of N-substituted tertiary 1,6-diMe-DHQ right after the flash the formation of intermediate particle with spectrum close to spectrum of particle B is observed (λ_{max} 490 nm). This particle is destructed by reaction of the first order with rate constant close to k_2 for corresponding secondary 6-Me-DHQ. We also determined the constants of rates of interactions between intermediate A and AcOH ($2\text{–}4 \times 10^8$ l·mole^{-1}·sec^{-1}) and intermediate B from secondary and tertiary DHQ and alkali (from $2{,}5 \times 10^7$ up to $2{,}5 \times 10^8$ l·mole^{-1}·sec^{-1} in dependence on substituents in aromatic cycle for secondary DHQ and 7×10^6 l·mole^{-1}·sec^{-1} for tertiary DHQ) [34].

On the base of carried out kinetic analysis of experimental data and analysis of reaction products the mechanism of photolysis process was proposed. It is presented in Scheme 15.1, where P are side products. Results obtained in neutral and acid mediums unambiguously testify that B is cation particle. Since addition occurs in accordance with Markovnikov's rule it should be corresponding carbo-cation. In the case of tertiary DHQ and for some secondary DHQ in water solutions the formation of carbo-cation B is observed already in nanosecond time diapason.

Kinetic isotope effects (KIE) of reactions (1) and (2) measured in MeOD confirm proposed mechanism [33]. KIE is equal to 2 for k_1 that is characteristic for reactions of proton transfer by A-S_E2 mechanism, and to 1,2–1,3 for k_2 that is characteristic for reactions of nucleophilic connection of solvent to carbo-cation [36]. In accordance with proposed mechanism the introduction of electron-donor substituents into aromatic nucleus stabilizers carbo-cation [34]. For series of compounds with substituent in position 6 this effect reveals more clearly than for compounds with substituent in position 8 (in MeOH k_2 = 80, 1830 and 2300 sec^{-1} for 6-EtO-DHQ, 8-MeO-DHQ and 6-H-DHQ accordingly).

Observing for secondary DHQ like titration curve concentration dependence of k_1^{eff} on alkali concentration is a consequence of reversibility of A into B transformation. Reactions (-1) in the case of secondary DHQ and (2') in the case of tertiary DHQ explain weak dependence of quantum yield of reaction on alkali concentration at its low concentrations. Introduction into scheme of interaction between A and alkali (reaction (3)) is necessary for adequate reflection of observed dependence of k_1^{eff} on alkali concentration for 8-MeO-DHQ [26]. In the case of compounds with substituents in position 6 this reaction doesn't proceed to a noticeable extent.

Particularity of mechanism of photo-induced addition of water and alcohols to DHQ is different ways of carbo-cation formation for secondary and tertiary DHQ [37]. Received experimental data testify that structure of A should be corresponding cycle orto-quinomethanimine as it is shown in Scheme 1.

This intermediate is generated under photolysis of secondary DHQ via double transfer of proton in excited singlet state with participation of molecules of H_2O or MeOH, and proton from solvent is transferred to C(3) atom, and from N–H bond to solvent. Presence of H_2O and MeOH is necessary condition for proceeding of this reaction. Thermal inverse transformation of A into initial DHQ is hindered that leads to long life times of this particle. Formation of this intermediate explains high values of pK_a = 9,4–10,4 received for carbo-cations generated from secondary DHQ under photolysis [38]. Unfortunately, in work [38] the received values of pK_a were mistakenly assigned to acid-basic equilibrium aminil radical–cation radical, since the authors thought that photolysis of DHQ occurred just as in hydrocarbon mediums.

It is obvious, that orto-quinomethanimine structure A is impossible for tertiary DHQ. For these DHQ the primary product of photo-chemical reaction is carbo-cation which is formed as a result of proton transfer from solvent to position C(3) of excited molecule of DHQ [37]. Carbo-cation independently on DHQ structure and way of its formation participates in reaction with nucleophilic solvent and forms adduct. In contrast to photochemical stage of reaction the necessary condition of which is the presence of H_2O and MeOH also the other alcohols interact with carbo-cation. Constant of rate of this reaction is determined by the presence of electron-donor substituents in aromatic cycle and by nucleophilisity of solvent [34]. Tertiary DHQ are more sensitive to transition from alcohols to water (for 6-Me-DHQ k_2 = 22 and 600 sec^{-1}, and for 1,6-diMe-DHQ k_2 = 13 and 790 sec^{-1} in H_2O and MeOH accordingly).

3. Reactivity of Carbo-Cations from DHQ in Relation to Nucleophiles

3.1. Reactivity of Anions in Relation to Carbo-Cations Generated under Photolysis of 1,2-Dihydroquinolines

Under photolysis of substituted DHQ the precursor of final adduct observed by method of impulse photolysis is carbo-cation. Carbo-cations are active intermediate particles of many organic reactions and their studying represents great interest (see reviews [39-41]).

Particularities of structure of carbo-cations formed under DHQ photolysis in water and MeOH lead to their large life-times (0,5-150 msec), high values of pK_a for carbo-cations from secondary DHQ (9,4-10,4) and high values of E^A for reaction of carbo-cation destruction (31–52 kJ/mole) [34]. Simplicity of registration by absorption spectra (λ_{Max} 480–500 nm) of these carbo-cations and also the absence of problem of contact ion pair formation under their generation allow investigating these intermediated in details.

As it was shown earlier, in alkali mediums there were differences in mechanisms of DHQ reactions photolysis for secondary and tertiary DHQ: in the case of secondary DHQ carbo-cation was not formed, whereas in the case of tertiary DHQ carbo-cation was observed even in alkali solutions, its life-time was reduced linearly in dependence on alkali concentration and other intermediates were not formed [42].

Received experimental data were interpreted from the point of view of existence of two resonance structures of formed cation: carbo-cation and orto-quinomethanimine. Increase of life time of the first intermediate particle in the case of secondary DHQ is explained by reverse reaction of cation in orto-quinomethanimine form with KOH (k_{-1}) (Scheme 2), i.e. by shift of acid-basic equilibrium in alkali medium to the side of formation of non-protonated form.

R^1 = H; MeOH, k_1; KOH, k_{-1}; R^1 = Me; KOH, k_q; OH; N; R^1

Scheme 2.

From the other hand, quenchering of cation from tertiary DHQ by alkali was interpreted by its interaction in the form of carbo-cation with OH^- with formation of corresponding hydroxyl-adduct (Scheme 2). In the case of secondary DHQ the hydroxyl-adduct in MeOH was not experimentally found. Obtained result, firstly, confirms dual reactivity of studied carbo-cations and, secondly, shows that rate constant of addition of hydroxyl ion to carbo-cation in the case of secondary DHQ is at least in order lower than for tertiary DHQ [42].

Studying of particularities of interaction of carbo-cation resulted from tertiary DHQ under photolysis with azide ion (N_3^-) which is standard nucleophile for estimation of reactivity of carbo-cations showed that carbo-cation from 1,6-diMe-DHQ reacted in methanol with N_3^- with constants of rates lower than diffusion controlled ones [27]. Particularities of interaction of studied carbo-cations with N_3^- in methanole are determined firstly by competition of reaction with MeOH, secondly by strong salt effect which is revealed in reduction of biomolecular constant of rate of interaction under increase of N_3^- concentration and its independence on this concentration at constant ion force at the expense of added $NaClO_4$. From the data on dependence of constant of rate of interaction between N_3^- and carbo-cation calculated at various ion forces of solution (μ) the values of this rate constant at μ = 0 and 0,8 were determined and turned to be equal to $2,0 \times 10^8$ and $4,0 \times 10^6$ l $mole^{-1}sec^{-1}$

accordingly at 18°C [27]. Activation barrier of reaction E^a = 20 ± 1 kJ/mole practically doesn't depend in ion force of solution [27].

Studied carbo-cation is relatively stable and constant of its destruction rate in MeOH is changed from 300 up to 3000sec^{-1} in temperature interval 0-50°C. According to McClelland et. al. [43] classification 5reaction of such carbo-cation with azide ion should proceed with lower rate constant than for reactions controlled by diffusion. Presented in given work experimental results confirm this prediction. By reaction ability in relation to azide ion (k_{Az}^0 = (1,1–3,5) × 10^8 l $mole^{-1}$·sec^{-1} in the same temperature interval) carbo-cation from 1,6-diMeDHQ is close to 3-Me-4'-MeO- and 4-Me-4'-MeO-triphenylmethyl cations for which rate constants of reaction with azide ion in water are k_{Az}^0 = 6,0 × 10^8 and 3,8 × 10^8 l $mole^{-1}$ sec^{-1} (20°C) [43] accordingly. However for these cations the values of rate constants of interaction with water are significantly higher than for 1,6-diMe-DHQ (k_w = 1100 and 570 sec^{-1} for triphenylmethyl cations [43], whereas 1,6-diMe-DHQ k_w = 13 sec^{-1} [34]). We should note that reaction rate constant for investigated carbo-cation with azide ion is approximately in 35 times higher that rate constant of this carbo-cation reaction with hydroxyl ion in MeOH (k_{OH} = 7 × 10^6 l mol^{-1} sec^{-1} at 20°C) [42]. Ratio between these rates constants is lower than was observed earlier for triphenylmethyl carbo-cations in water for which $lg(k_{Az}/k_{OH}) \approx 2,8$ [44]. Observed differences in ratios of rates constants for triphenylmethyl carbo-cations and carbo-cations from DHQ are firstly connected with differences in their structure (in [43] they notice that for each structural class of carbo-cations there is its own correlation dependence between rates constants, although they are qualitatively analogous), and secondly, presented in given work rates constants were measured in methanol , and in [43, 44] they were measured in water. Change of solvation sphere of carbo-cations also may influence on their reactivity in reaction with nucleophillic ions.

Temperature dependence of k_{Az} is additional proof of activation control of reaction, and activation energy of this reaction is significantly lower than for reactions of investigated carbo-cation with MeOH and H_2O (E^a(MeOH) = 32 and $E^a(H_2O)$ = 50 kJ/mole) [34]. Unfortunately, there are too little data in literature on temperature dependences of constants of rates of carbo-cations interactions with nucleophiles. Moreover these dependences were obtained by indirect methods in experiments on solvolysis and they determined the differences in activation energies of reactions of carbo-cations with solvent and with N_3^-. Ingold with collaborators found under di-n-tolylmethylchloride hydrolysis that activation energy for reaction of corresponding carbo-cation with water was by 16,8 kJ/mole lower than for reaction of the same carbo-cation with N_3^- [45]. Banton with collaborators determined the difference in activation energies $E^a(N_3^-)$ – E^a(MeOH) = 8,4–12,6 kJ/mole for carbo-cation resulted from camphene hydrochloride solvolysis [46]. In other words, in these cases in spite of the fact that reaction with azide ion is faster than reaction with nucleophillic solvent the activation energy for it is higher. This question was considered by Ingold [47] and he assumed that in the case of carbo-cations presenting in solvate sphere of nucleophilic solvent reaction with solvent occurred with lower activation energy than reaction with anion for which additional energy was necessary to penetrate through this solvate sphere. The question is about why reactions of investigated carbo-cation with MeOH and water proceed with higher activation energy than with N_3^- will be discussed below.

3.2. Relative Reactivity of Alcohols and Water in Relation to Carbo-Cations Generated under Dihydroquinolines Photolysis

Why does not selectivity of addition reactions correspond to observed rates constants? Investigation of dependences of rate constant of investigated carbo-cations destruction in mixtures R'OH—R"OH, where R' = Me, Et and Pr^n, and R" = H and Me, on mixture structure allowed determination of relative reactivity of these nucleophiles in relation to carbo-cations and making conclusions about the mechanism of their addition reaction. In water neutral solutions under photolysis of secondary DHQ the carbo-cations formation already in nanosecond time diapason is observed. In neutral alcohol solutions the consecutive formation of cycle orto-quinomethanimime and cation is observed. In acidulous AcOH ([AcOH] >1 × 10^{-5} mole l^{-1}) alcohol solutions life time of the first intermediate product cycle orto-quinomethanimine is sharply reduced and under registration of intermediate particles in millisecond time scalethe formation of only carbo-cation is observed destruction of which occurs by reaction of pseudo-first order. In the case of tertiary DHQ formation of carbo-cation is observed already in nanosecond time interval even in weak alkali solutions and their destruction occur by reaction of the first order in millisecond time diapason. That is why when studying reactivity of carbo-cartion from secondary DHQ in relation to alcohols the experiments were carried out in solutions acidulated by AcOH [48].

In Table 1 measured values of constants of pseudo-first order destruction reactions of carbo-cations (k_2) formed under impulse photolysis of various DHQ in alcohols and water are presented.

Table 1. The ratio of observed rates constants of carbo-cation destruction in pure solvents ($a = k_2^{R'OH}/k_2^{R''OH}$), selectivity (S) in mixture of solvents (R'OH + R"OH) and parameters calculated by approximation of experimental data with the use of equations (7) and (8): constant of solvation equilibrium (K), ratio of equilibrium constants of the first stage of reaction, i.e. addition of ROH for various solvents (b = K'/K") and ratio of rate constants of deprotonation of intermediate ion for pure components of mixture ($c = k'_{-H}/k''_{-H}$).

DHQ	System R'OH—R"OH	a	S	K	b	c
6-Me-DHQ	MeOH—H_2O	30	1,42	1,3	3,8	7,9
	EtOH—H_2O	26,5	0,56	2,4	2,2	11,8
	Pr^nOH—H_2O	26,5	0,46	2,5	2,1	12,4
	Pr^nOH—MeOH	0,88	0,41	0,9	0,7	1,18
	Pr^iOH—MeOH	-	-	0,8	-	1,5
6-EtO-DHQ	MeOH—H_2O	22	1,5	0,9	3,0	7
	EtOH—H_2O	20	0,56	2,0	2,2	8,9
	Pr^iOH—MeOH	-	-	1,0	-	1,6
1,6-diMe-DHQ	MeOH—H_2O	60	2	1,3	5,2	11,4
	EtOH—H_2O	65,4	1	1,8	3,6	18
	Pr^nOH—MeOH	1,16	0,41	0,95	0,8	1,5
	Pr^iOH—MeOH	-	-	0,8	-	1,5

Since under photo-excitation of DHQ in pure EtOH and Pr^nOH the carbo-cation precursor and consequently carbo-cation itself are not formed, then values of k_2 for these alcohols were obtained by extrapolation of values of rate constants of carbo-cation destruction in mixtures of these alcohols with water and MeOH (k_s) (Figure 1a), and for Pr^nOH one and the same value in mixtures with both solvents was obtained.

Values of k_2 for various DHQ are changed in accordance with electron-donor ability of substituents. For given DHQ values of k_2^{ROH} in alcohols are close to each other, but more than in 15-20 times exceed the rate constant in water (Table 1), that corresponds to increase of nucleophilicity of alcohols in comparison with water.

When carrying out the reaction in mixture of two nucleophylic solvents the ratio of adducts concentrations formed in mixtures should correspond to structure of mixture according to equation (1):

$$[P']/[P''] = \mathbf{S}[R'OH]/[R''OH], \quad (1)$$

where S = k'/k" – selectivity of reaction which in the case of two parallel reactions is determined by ratio of rate constants, i.e. linear dependence between adducts structure and mixture composition should be observed.

As it is obvious from Figure 2 the dependence (1) is actually observed for mixture of solvents in the case of DHQ photolysis although not in the whole region of concentrations change, especially in the case of H_2O mixtures with EtOH and Pr^nOH. In this case linearity is observed up to H_2O concentration not higher than 50%. At higher water concentrations predominant formation of hydroxy-adduct was observed.

In Table 1 experimental values of parameter $a = k_2^{R'OH}/k_2^{R''OH}$ and selectivity S measured from experimental data on structure of reaction products for various DHQ in various mixtures of nucleophilic solvents are compared. It is obvious from Table 1 that photolysis selectivity in mixtures is significantly lower than one may expect basin on ratio between rate constants of carbo-cations in pure solvents (see curves 2 and 4, Table 1). As the results of products analysis testify MeOH is only in 2-1,4 times more active than water, and EtOH and Pr^nOH even less active than water. The last observation contradicts higher nucleophility of these solvents in comparison with water and it is connected with steric hindrances under formation of adducts with these alcohols. actually, in ^1H NMR spectra of EtOH and Pr^nOH adducts formation of two isomers was observed that was conditioned by hindered rotation of added alkoxy-group. MeOH is in 2,5 times more active in addition reaction than Pr^nOH (Figure 1, curve 3).

The synergetic effect is observed in alcohols mixture, when carbo-cation destruction rate constant k_s is higher than k_2 of each alcohol separately (Figure 1). The fact of k_s increase more than by order in water mixtures with Pr^iOH which is inactive in addition reaction is even more surprising (Figure 1). Thus, we obtain, that in the case of carbo-cations generated under DHQ photolysis measured selectivity is significantly differed from ratio of rate constant of carbo-cations destruction measured by method of impulse photolysis in individual solvents.

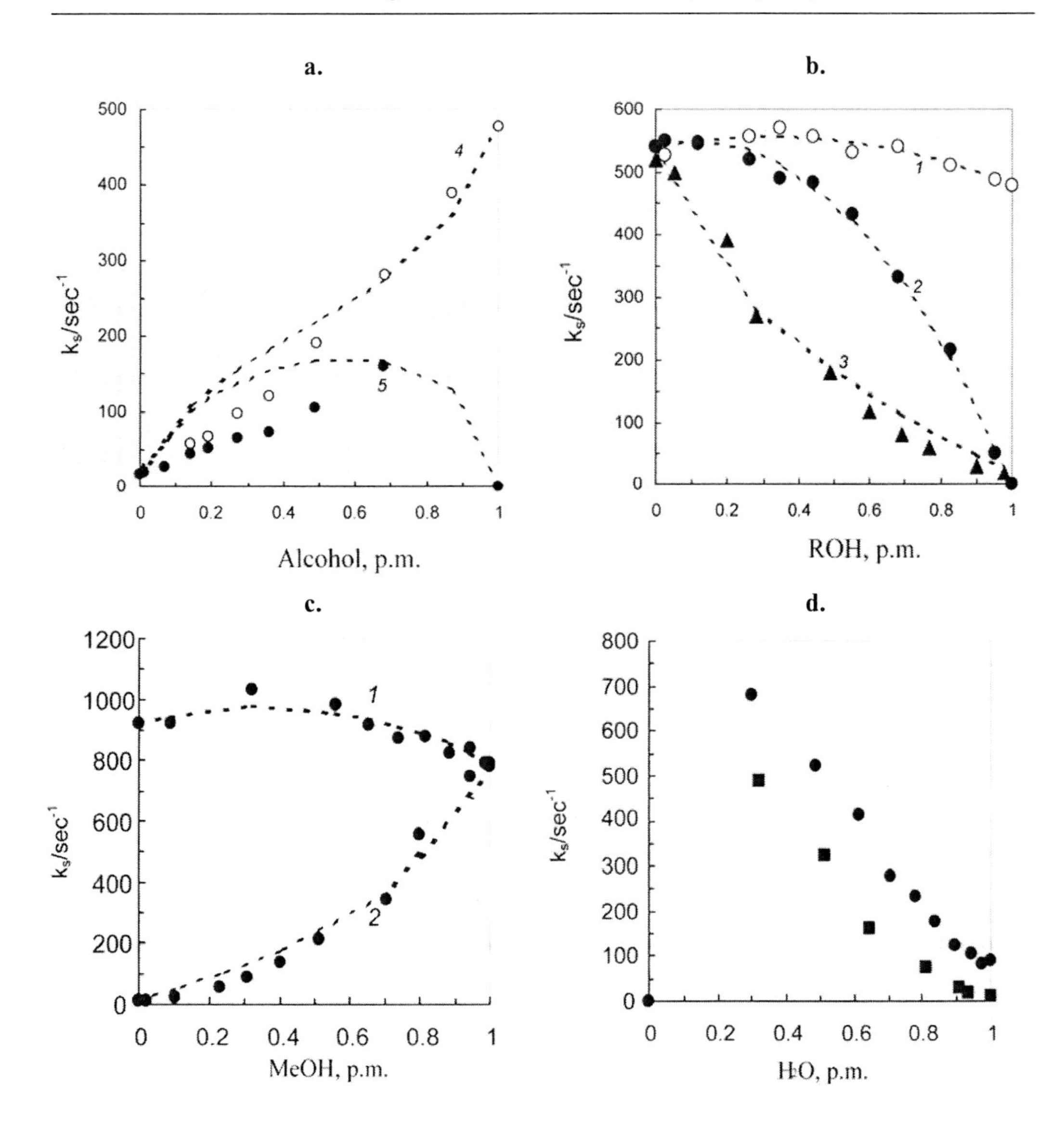

Figure 1. Dependence of rate constant of carbo-cation destruction k_s on mixture structure: (a) and (b) for 6-Me-DHQ in mixtures MeOH—Pr^nOH (1), MeOH—Pr^iOH (2), MeOH—H_2O (3), H_2O—Pr^nOH (4) and H_2O—Pr^iOH (5), concentration on abscissa axis corresponds to the second component of mixture; (c) for compound 1,6-diMe-DHQ in mixtures MeOH—Pr^nOH (1), MeOH—H_2O (2); (d) for compound 6-H-DHQ (●) and 1,6-diMe-DHQ (■) in mixture H_2O—Pr^iOH. Points present experimental data and dotted curves present calculation by equations (7) and (8).

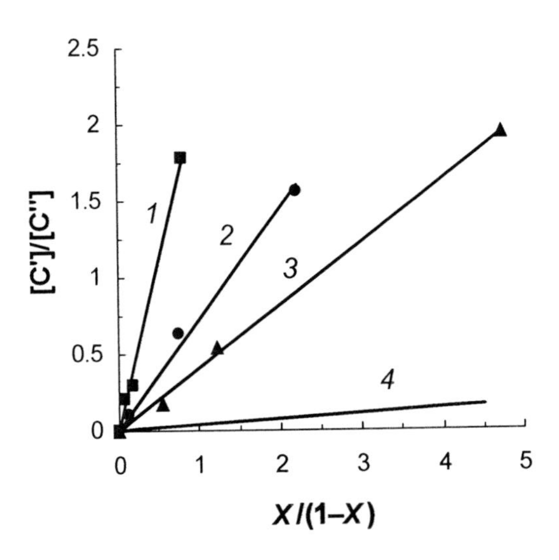

Figure 2. Dependence of structure of reaction products of water and alcohols photo-addition to 6-Me-DHQ on solvent structure R'OH + R"OH (the first solvent is that one which mole part is X): (1) H_2O + PrOH, (2) H_2O + MeOH, (3) MeOH + PrOH, (4) dependence for mixture H_2O + MeOH, calculated for ratio $S = a = k_2^{R'OH}/k_2^{R''OH}$.

Received experimental data can't be explained by solvent polarity change in mixtures, since consideration of dependence of rate constant on solvent polarity in the view $\lg k_s \sim (1 - 1/\varepsilon)$ [49] in considered case gives correction for k_s lower than 5% in less polar solvent and that is why it can't significantly change constants ratio. This fact corresponds to theoretical conceptions [49] and experimental data [50] about the fact that in reactions of ion with molecules the dependence of reaction rate on solvent polarity is not big.

The question about solvent's nature and solvation influence on carbo-cations reactions kinetics is studied in literature in details. It is known, that in solvents mixtures ions are solvated predominantly by more polar solvent [47, 49]. In the case of the simplest scheme of consideration of predominant solvation by one of the mixture components (R"OH) [49] the composition of products linearly depends on solvent structure, and selectivity is determined by the expression (2):

$$S = k'/2Kk'', \qquad (2)$$

i.e. the factor 2K is appeared connected with predominant solvation by one of the solvent components. In this case the dependence of rate constant of carbo-cation destruction measured in mixtures should obey the equation (3):

$$k_s = 2k'(SX + (1 - X))/(SX + k'(1 - X)/k'') \qquad (3)$$

where X and (1-X) – mole parts of solvents R'OH and R"OH accordingly. However equation (3) doesn't describe experimental dependences of change of observed rate constant on mixture structure in the case of DHQ. Apparent contradiction between selectivity in mixtures and values of k_2 in individual solvents and also acceleration of destruction of carbo-cation in

Pr^iOH may be explained if one presents the mechanism of carbo-cation interaction with solvent as a consecution of two reactions:

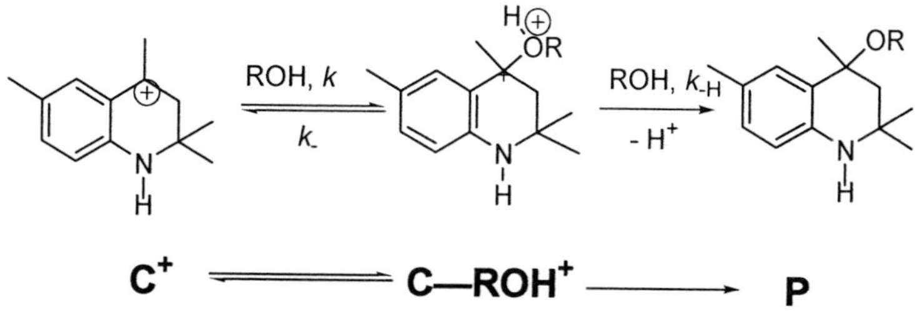

ROH, k / k_- ; ROH, k_{-H} / $- H^+$

$C^+ \rightleftharpoons C\text{—}ROH^+ \longrightarrow P$

Scheme 3.

i.e. if one separates stages of addition of ROH and elimination of proton. For mixture of solvents (R'OH + R"OH) this scheme with consideration of predominant solvation of carbo-cation R"OH will be as follows:

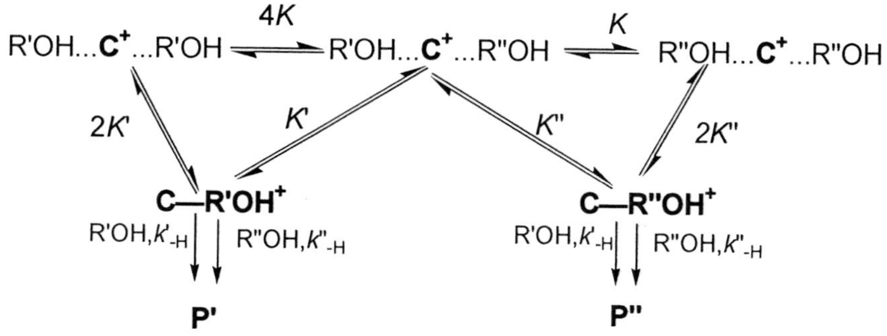

R'OH...C^+...R'OH ⇌ ($4K$) R'OH...C^+...R"OH ⇌ (K) R"OH...C^+...R"OH

$2K'$, K', K'', $2K''$

C—R'OH$^+$ → (R'OH,k'_{-H}; R"OH,k''_{-H}) P'

C—R"OH$^+$ → (R'OH,k'_{-H}; R"OH,k''_{-H}) P"

Scheme 4.

Composition of products in this case will be determined by equation analogous to equation (2) in which rate constants k' and k" are substituted by equilibrium constants K' and K":

$$S = K'/2KK'', \tag{4}$$

where K' and K" – equilibrium constants of the first stage for R'OH and R"OH accordingly, i.e. reaction selectivity is determined by the first equilibrium stage of process. We should note that reaction selectivity weakly depends on investigated DHQ and is determined mainly by nature of components forming the mixture (Table 1). Constants of carbo-cation destruction rate according to Scheme 4 depends on rate constant of the second stage – deprotonation of formed intermediate ion, i.e. on affinity to proton of solvents components. That is why this reaction is significantly accelerated in alcohols for which affinity ti proton if significantly higher than for water [51].

It is clear from Scheme 4 why addition of Pr^iOH into water and MeOH leads to acceleration of carbo-cation destruction. It becomes possible since rate constant of proton transfer on solvent (k_{-H}) is higher for Pr^iOH than for water and MeOH [51], i.e. Pr^iOH doesn't participate directly in adduct formation but nevertheless accelerates this reaction.

Accurate solution of system of differential equations corresponding to reactions totality presented in Scheme 4 is biexponential. However observed kinetics of carbo-cation destruction in experiments of impulse photolysis is strictly monoexponential that testifies to realization of one of absolute cases considered below. Under quantitative consideration of this scheme for solvent concentration at stage of solvation equilibrium the mole parts X and (1 – X) were used for R'OH and R"OH accordingly. For reaction of proton elimination from intermediate cation they used molar concentration of solvent [R'OH] = $[R'OH]_0 v$ and [R"OH] = $[R''OH]_0(1 - v)$, where v and (1 – v) are volume parts of components R'OH and R"OH accordingly, and $[R'OH]_0$ and $[R''OH]_0$ are the concentrations of these components in pure solvent. Then, if we use the values of rate constants of pseudo-first order in pure solvent k'_{-H} и k''_{-H} we obtain $k'^{mix}_{-H} = k'_{-H}v$, and $k''^{mix}_{-H} = k''_{-H}(1 - v)$.

If one assumes quasi-stationarity by intermediate cations $C—R'OH^+$ and $C—R''OH^+$ and independence of k_{-H} on $C—ROH^+$ that signifies that this rate constant is determined mainly by affinity to proton of solvent one may obtain simple expressions for observed in mixtures rate constants for two absolute cases.

If:

$$k_-^i << k_{-H}^i, \tag{5}$$

i.e. formed intermediate cation $C—ROH^+$ is stable enough and it is faster transformed into final adduct than destructed with formation of initial components, selectivity at conditions of predominant solvation of R"OH comes to previous variant and is described by expression (2), and dependence of k_s on solvent structure is described by expression (3).

If formed cation $C—ROH^+$ is unstable and is decomposed into initial components, i.e. the condition (6) is observed:

$$k^i_- >> k^i_{-H}, \tag{6}$$

then dependence of k_s on structure of solvent is described by expression (7).

$$k_s = k_2^{R''OH}(av + 2KS(1 - v)(SX + (1 - X))/S(2K(1 - X) + X) \tag{7}$$

where parameters a, S and K were determined above. Experimental data received by dependence of k_s on solvent structure were approximated in accordance with equation (7) with the use of experimental values of a and S and solvation constant K as variable parameter. If we determined K we might calculate the ratio of constants of rate and equilibrium for various DHQ in various solvents:

$$b = K'/K'' = 2SK \text{ and } c = k'_{-H}/k''_{-H} = a/b.$$

Approximation was carried for those mixtures in which adducts structure linearly depended on solvent structure. In the case of alcohols mixtures equation (7) is observed in the whole region of solvent structure change. For mixture of alcohols with water equation (7) is observed if mole part of water is lower than 0,6 for secondary DHQ. In the case of tertiary DHQ due to instability of hydroxy-derivative the error of determination of adducts structure was significantly higher, so these data should be considered as evaluating. Obtained values of parameters K, b and c are presented in Table 1, and calculated dependences of k_s on solvent structure in Figure 1. It is obvious from them that equation (7) satisfactorily describes experimental data in the region of linear dependence of adducts structure of solvent composition. Deviations from experimental dependences in the region of high water concentrations testify to the fact that prevailing solvation of water is stronger than used model proposed (Scheme 4).

As it is obvious from Table 1 equilibrium constant of solvation K depends on solvent components to a greater extent than on nature of studied DHQ. Only in the case of 6-EtO-DHQ the inversion for mixture MeOH—H_2O is observed in comparison with other DHQ, namely: for this compound carbo-cation solvation by both solvents occurs practically to equal extent. This fact may be connected with higher hydrophobicity of ethoxy group that is reflected also on the shape of dependence of k_s on X which in the case of 6-EtO-DHQ is less concave in the region of higher water concentrations, than for other compounds in this mixture. For alcohols mixture K is close to 1, i.e. predominant solvation of carbo-cation by one of the components in these mixtures is practically absent that corresponds to ε values proximity for these solvents.

Parameter b reflects mainly relative electron-donor ability of mixture components in relation to studied carbo-cation, and also possible steric and orientation factors influencing on adduct formation. As it is obvious from obtained data this parameter in 2-4 times higher for alcohols than for water that corresponds to increase of electron-donor ability under transition from water to alcohols [51]. However, the fact that among alcohols the maximum value was observed for methanol obviously characterized the less steric hindrances under formation of adduct with MeOH in comparison with other alcohols.

Parameter **c** is determined by relative affinity to proton and for alcohols this parameter is approximately by order higher than for water, and in the case of alcohols it is increased in the raw MeOH < $Pr^nOH \approx$ EtOH, that also corresponds to literature data [51]. Here we should note that parameter **c** presented in Table 1 is equal to ratio of constants of pseudo-first order rate of intermediate cation deprotonation in pure solvents. If one considers ratio of bimolecular constants using concentrations of components in pure solvents then the raw of alcohols will be MeOH < EtOH < Pr^nOH.

The discussed Scheme 3 is not original. It was proposed by Bannett [52] for olefins reactions with acids in water mediums and is practically always used when carbo-cations reactions with water and alcohols are discussed (see [50]). However, it was not used in such meaning as it did in given work, since for investigated earlier carbo-cations the selectivity of addition reaction of solvent in mixture [50, 53] didn't conflict with values of carbo-cations destruction rate constants measured by direct methods recently [43, 54]. This fact means that for cations which are discussed in [43, 50, 53, 54] either the equation (5) is observed, or formation of intermediate cation doesn't occur, i.e. the basic catalysis of reaction is realized as it was proposed for very active cations in [53]. In these cases products structure should correspond to observed rate constants. In contrast to [50, 52] in Schemes 3 and 4 the reaction 2 seems to be irreversible, since under DHQ photolysis the quantitative addition of solvent occurs and reverse reaction may be neglected.

Concerning solvation effects Ingold [47] noticed that in the presence of enough amount of water primary formed carbene ion is covalently bonded with one of the solvating water molecules and doesn't depend on structure of prevailed medium mass. Although this reasonings relate to mixtures of water with acetone they obviously reflect the situation in water-alcohol solutions. Weak dependence of rate constant of the first order for carbo-cation destruction formed under photolysis of triarylacetonitriles on water concentration in mixture with acetonitrile was noticed by M.G. Kuz'min et. al. [55]. They explained this fact by formation of water solvate shell. In [50] they noticed the increase of biomolecular constant of rate of reaction between water and carbo-cations under water concentration decrease. In fact this is the same effect, since for reactions proceeded between particles in solvate shell biomolecular rate constant is uncertain, because usage of solvent-reagent volume concentration in given case is not correct. Our experimental data also testify that in the case of water-alcohol solutions the carbo-cation is predominantly solvated by water.

3.3. Particularities of Carbo-Cation Reactions in Mixtures of Solvents Containing Isopropanol

Studying of reaction of carbo-cations destruction in mixtures containing Pr^iOH which due to steric hindrances doesn't form adducts with studied carbo-cations is of great interest for revealing of role of proton elimination reaction. As it is obvious from Figure 1 (a and d) constant of carbo-cation destruction rate k_s is increased more than in order under decrease of water mole part down to 0,3. Unfortunately, at further increase of mole part of Pr^iOH quantum yield of carbo-cations under DHQ photolysis is sharply decreased and measuring of k_s becomes impossible. For mixtures of Pr^iOH with MeOH under reduction of MeOH content firstly the small rise of k_s, and then decrease down to 0 are observed (Figure 1).

For mixtures containing Pr^iOH the scheme of reaction 4 is simplified, since in this case we should consider formation of only one intermediate product **C**–R"OH^+ with further transfer of proton on both components of mixture (Scheme 5).

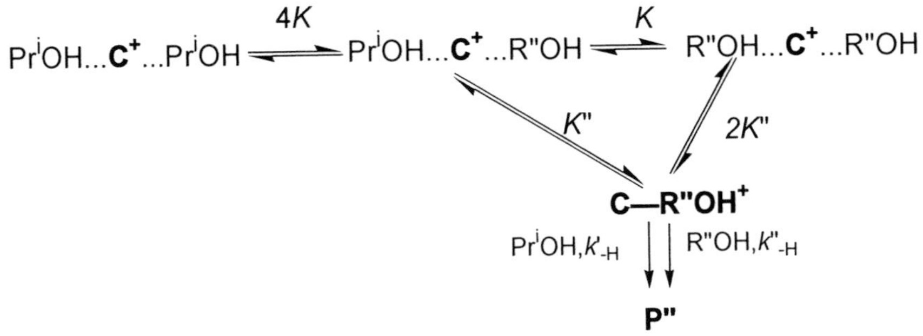

Pr^iOH...**C^+**...Pr^iOH ⇌(4*K*) Pr^iOH...**C^+**...R"OH ⇌(*K*) R"OH...**C^+**...R"OH

Pr^iOH...**C^+**...R"OH ⇌(*K*") **C—R"OH^+** ⇌(2*K*") R"OH...**C^+**...R"OH

C—R"OH^+ → (Pr^iOH, K'_{-H}; R"OH, $K"_{-H}$) **P"**

Scheme 5.

Corresponding expression for observed constant of carbo-cation destruction rate in mixture is as follows:

$$k_s = 2k_2^{R"OH} K(cv + (1 - v))(1 - X)/(X + 2K(1 - X)), \tag{8}$$

where $2k_2^{R"OH} = 2K"k"_{-H}$ – constant of carbo-cation destruction rate in pure solvent, $c = k'_{-H}/k"_{-H}$ and K – constant of solvation equilibrium. Equation (8) adequately reflects the dependence of k_s on mixture composition Pr^iOH–MeOH. Values of K and c obtained as a result of approximation of experimental curves of dependence of k_s on mixture composition are presented in Table 1. Comparison of these values with values of K and c for mixtures of Pr^nOH with MeOH shows their proximity and value of K is somewhat lower, and c is higher for Pr^iOH than for Pr^nOH.

Unfortunately such approximation can't be realized for mixtures of Pr^iOH with water, since as it was shown above at mixture structure where reactions of carbo-cation may be observed the solvation of carbo-cation by water is stronger than Scheme 5 proposed. In Figure 1 the calculated curve of k_s dependence on mixture structure Pr^iOH—H_2O for 6-Me-DHQ compound is presented with proposition that parameters K and c correspond to those ones obtained for mixture Pr^nOH—H_2O. The fact that maximum of k_s corresponds to mole part of H_2O 0,3 and is practically equal to experimentally observed value is remarkable. Although at higher water concentrations there is no quantitative coincidence we can make some qualitative conclusions on the base of these data. At mole part of water higher than 0,8 the both experimental and calculated dependences for Pr^nOH and Pr^iOH coincide, that testifies to the fact that at these water concentrations addition reaction is realized only with water, and small reaction acceleration under alcohols addition is caused by evolving of alcohols into reaction of intermediate cation deprotonation. At further reduction of water concentration the curves for Pr^iOH and Pr^nOH are diverged, since Pr^nOH begines to participate also in reaction of addition, and Pr^iOH reacts only with intermediate cation.

The results of measuring of activation energies of reaction of carbo-cation destruction in water (E^a_{H2O}) and in mixture with Pr^iOH with mole part of water 0,3 (E^a_{mix}) turned to be very interesting. Activation energy in water for all investigated DHQ is equal to 41-50 kJ/mole, and E^a_{mix} = 30-35 kJ/mole, i.e. under transition form pure water to mixture Pr^iOH—H_2O activation energy of carbo-cation destruction reaction is reduced by 11-15 kJ/mole. Since reaction of intermediate cation formation for these two solvents is common, then difference in activation energies reflects the difference in reaction activation energies of its deprotonation by water and Pr^iOH. Thus, deprotonation reaction of intermediate cation by water requires significantly higher activation energy than Pr^iOH. The fact that activation energies of carbo-cation destruction reaction for various DHQ in mixture Pr^iOH—H_2O are close to activation energies of carbo-cation destruction in methanol which is equal to 30-35 kJ/mole is also noticeable.

3.4. Structure of Carbo-Cations and Activation Barrier of Reactions with Nucleophiles

Studied carbo-cations have long life times and their reactions with nucleophilic agents proceed with significant activation energy. And in contrast to carbo-cations studied in literature activation energy of reaction with ion nucleophils is significantly lower than with molecular ones. Quantum-chemical calculations of carbo-cations structure from DHQ carried out by PM-3 in Gaussian-98 program showed that positive charge in them was strongly delocalized, and the charge on nitrogen atom was equal to ~ +0,34, and on carbon atom in position 4 to ~ +0,13. Calculations also showed that under gradual lengthening of bonds C^9—C^4 and C^{10}—N with optimization of structure at each step the positive charge was gradually shifted from nitrogen atom on carbon atom in position 4. Structure with zero charge on nitrogen atom and positive charge on C^4 corresponding to structure of classical carbo-cation was higher in energy almost by 50 kJ/mole than main structure. Thus, high barrier of reaction with nucleophils is caused first of all by necessity of activation of carbo-cation.

The contribution into stability of investigated carbo-cation is also made by steric factor since formation of optimal reagents configuration corresponding to reaction is hindered due to methyl substituents in positions 2 and 4. Unfortunately we were not succeeded in estimation of this contribution since DHQ without substituents in hetero-cycle were too unstable. Particularities of addition of molecular nucleophils to carbo-cations from DHQ allow us assuming that resulted from addition of ROH to carbo-cation intermediate ion is extremely unstable and by energy it is situated close to transition point on the surface of potential energy (Figure 3), and its transformation into final product requires additional activation energy as it is obvious from data on measuring of activation energy in water and mixtures Pr^iOH—H_2O. This fact makes additional contribution into reaction activation energy, increases it in comparison with azide ion even if activation energy of reaction with azide ion is higher than activation energy of the first stage of proton solvent addition to carbo-cation (see Figure 3a and c). The same fact tells reactions of DHQ carbo-cations from traditional carbene iones for which activation energy of the second stage of reaction is small (Figure 3b, block curve), and in the case of very active unstable carbo-cations reaction proceeds in one stage (Figure 3b, dotted curve).

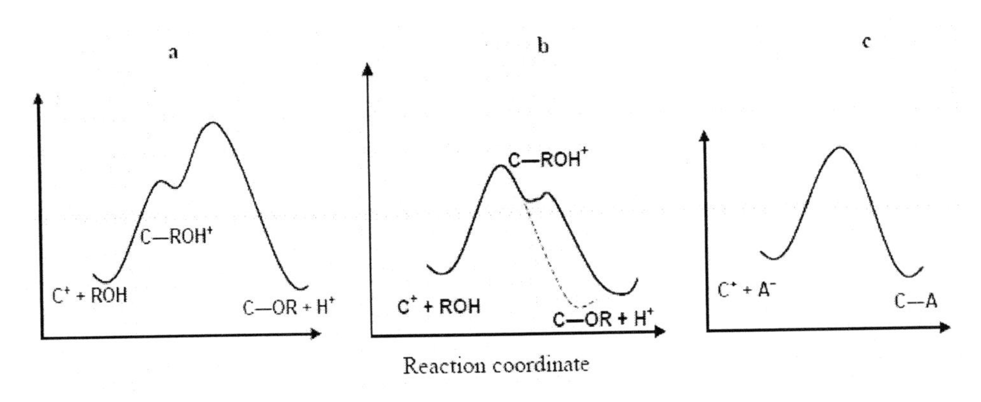

Figure 3. Schematic presentation of addition reaction of molecular (a) and (b) and ion (c) nucleophils in the case of carbo-cation from DHQ (a) and (c) and classical carbo-cation (b).

Thus, application of approaches to investigation of chemical reactions mechanism developed under direction of Professor N.M. Emanuel allowed revealing of the novel class of reactions for 2,2,4-trimethyl-1,2-dihydroquinolines in proton solvents, namely the photo-induced addition of water and alcohols to double bond of DHQ, establishing of mechanism of these reactions and determining of spectral-kinetic characteristics of intermediate compounds. Special attention was paid to key intermediate compounds carbene ions formed under DHQ photolysis which together with properties were common for carbo-cations resulted from solvolysis of many organic substances and had a number of distinctive features. Detailed studying of kinetic regularities of reactions of carbo-cation from DHQ with ion and molecular nucleophils in dependence on structure of solvent showed that their relative stability was determined by resonance stabilization of carbo-cation and steric hindrances under adducts formation. The important role of solvent in these reactions was revealed. these investigations contribute into understanding of carbo-cation interactions with nucleophils.

REFERENCES

[1] T.D. Nekipelova, A.B. Gagarina, *Dokl. AN SSSR*, 231, 392 (1976) *(in Russian).*

[2] T.D. Nekipelova, A.B. Gagarina, N.M. Emanuel, *Dokl. AN SSSR*, 238, 392, (1978) *(in Russian).*

[3] T.D. Nekipelova, A.B. Gagarina, N.M. Emanuel, *Dokl. AN SSSR*, 238, 630, (1978) *(in Russian).*

[4] T.D. Nekipelova, A.B. Gagarina, N.M. Emanuel, *Izv. AN SSSR, Ser. Khim.*, 734 (1979) *(in Russian).*

[5] T.D. Nekipelova, A.B. Gagarina, *Dokl. AN SSSR*, 226, 125 (1976) *(in Russian).*

[6] T.D. Nekipelova, A.B. Gagarina, *Dokl. AN SSSR*, 226, 626 (1976) *(in Russian).*

[7] T.D. Nekipelova, A.B. Gagarina, *Neftekhimiya*, 22, 278 (1982) *(in Russian).*

[8] E. Knoevenagel, *Ber.*, 54, 1722 (1921).

[9] G. Reddelien, A. Thurn, *Ber.*, 65, 1511 (1932).

[10] W. . Pryor, T. Strickland, D.F. Church, *J. Am. Chem. Soc.*, 110, 2224 (1988).

[11] B.I. Gorbunov, Ya.A. Gurvich, I.P. Maslova, *Chemistry and technology of polymer materials*, Moscow: Khimiya (1981) *(in Russian).*

[12] Patent № 28078 Australia (1971); Patent № 3907507 USA (1973); Patent № 5–29473 Japan (1972).

[13] J. Pospisil, *Adv. Pol. Sci.*, 124, 87 (1995).

[14] J.U. Skaare, E. Solheim, *Xenobiotica*, 11, 649 (1979).

[15] J.U. Skaare, *Xenobiotica*, 11, 659 (1979).

[16] B. Lockhart, N. Bonhomme, A. Roger, G. Dorey, P. Casara, P. Lestage, *Eur. J. Pharm.*, 416, 59 (2001).

[17] T.B. Saxena, K.E. Zachariassen, L. Jorgensen, *Comparative Biochem., Physiol. C-Toxic., Pharm.*, 127, 1 (2000).

[18] S. Thorisson, F.D. Gunstone, R. Hardy, *Chem., Phys. Lipids*, 60, 263 (1992).

[19] A.M. Galzin, M. Delanaye, C. Hoornaert, G. McCort, S. E. O'Connor, *Eur. J. Pharm.*, 404, 361 (2000).

[20] A. Miralles, C. Ribas, P. V. Escriba, J. A. Garcia-Sevilla, *Pharm., Toxicol.*, 87, 269 (2000).

[21] H. Iwamura, H. Suzuki, Y. Ueda, T. Kaya, T. Inaba, *J. Pharm., Experim. Therap.*, 296, 420 (2001).

[22] F.I. Tarazi, N.S. Kula, K.H. Zhang, R.J. Baldessarini, *Neuropharmacology*, 39, 2133 (2000).

[23] S. Jaroch, P. Holscher, H. Rehwinkel, D. Sulzle, G. Burton, M. Hillmann, F. M. McDonald, *Bioorg., Med. Chem. Lett.*, 12, 2561 (2002).

[24] T.D. Nekipelova, L.N. Kurkovskaya, I.I. Levina, N.A. Klyuev, V.A. Kuz'min, *Izv. AN SSSR, Ser. Khim.*, 2072 (1999) *(in Russian).*

[25] T.D. Nekipelova, L.N. Kurkovskaya, I.I. Levina, V.S. Shishkov, V.A. Kuz'min, *Izv. AN SSSR, Ser. Khim.*, 647 (2001) *(in Russian).*

[26] T.D. Nekipelova, Yu.A. Ivanov, E.N. Khodot, V.S. Shishkov, *Kinetika i kataliz*, 43, 333 (2002) *(in Russian)*.

[27] T.D. Nekipelova, I.I. Levina, P.P. Levin, V.A. Kuz'min, *Izv. AN SSSR, Ser. Khim.*, 772 (2004) *(in Russian)*.

[28] E. L. Land, G. Porter, *Trans. Far. Soc.*, 59, 2026 (1963).

[29] T.D. Nekipelova, Ya.N. Malkin, V.A. Kuz'min, *Izv. AN SSSR, Ser. Khim.*, 80 (1980) *(in Russian)*.

[30] F.D. Gunstone, R.C. Mordi, S. Thorisson, J.C. Walton, R.A. Jackson, *J. Chem. Soc., Perkin Trans. 2*, 1955 (1991).

[31] L. Taimr, M. Prusikova, J. Pospisil, *Angew. Makromol. Chem.*, 190, 53 (1991).

[32] O.T. Kasaikina, A.B. Gagarina, Yu.A. Ivanov, E.G. Rozantzev, N.M. Emanuel, *Izv. AN SSSR, Ser. Khim.*, 2247 (1975) *(in Russian)*.

[33] T.D. Nekipelova, *Photochem., Photobiol., Sci.*, 1, 204 (2002).

[34] T.D. Nekipelova, V.A. Kuz'min, V.S. Shishkov, *Khimiya vysokikh energii*, 36, 212 (2002) *(in Russian)*.

[35] T.D. Nekipelova, *Int. J. Photoenergy*, 1, 25 (1999).

[36] R. Bell, *Proton in chemistry*, Moscow: Mir (1977) *(in Russian)*.

[37] T.D. Nekipelova, V.S. Shishkov, *Khimiya vysokikh energii*, 38, 355 (2004) *(in Russian)*.

[38] Ya.N. Malkin, N.O. Pirogov, M.V. Kopytina, V.N. Nosova, *Izv. AN SSSR, Ser. Khim.*, 1866 (1984) *(in Russian)*.

[39] C.D. Ritchie, *Acc. Chem Res.*, 5, 348 (1972).

[40] R.A. McClelland, *Tetrahedron*, 52, 6823 (1996).

[41] J.P. Richard, T.L. Amyes, M.M. Toteva, *Acc. Chem Res.*, 34, 981 (2001).

[42] T.D. Nekipelova, L.N. Kurkovskaya, I.I. Levina, *Izv. AN SSSR, Ser. Khim.*, 1899 (2002) *(in Russian)*.

[43] R.A. McClelland, V.M. Kanagasabapathy, N. Banait, S. Steenken, *J. Am. Chem. Soc.*, 113, 1009 (1991).

[44] C.D. Ritchie, *Can. J. Chem.*, 64, 2239 (1986).

[45] A.R. Hawdon, E.D. Hughes, C.K. Ingold, *J. Chem. Soc.*, 2499 (1952).

[46] C.A. Bunton, T.W. Del Pesco, A.M. Dunlop, K.-H. Yang, *J. Org. Chem.*, 36, 887 (1971).

[47] K. Ingold, *Theoretical bases of organic chemistry*, Moscow: Mir (1973) *(in Russian)*.

[48] T.D. Nekipelova, I.I. Levina, V.S. Shishkov, *Kinetika i kataliz*, 45, 28 (2004) *(in Russian)*.

[49] J. Gordon, *Organic chemistry of electrolyte solutions*, Moscow: Mir (1979) *(in Russian)*.

[50] J.P. Richard, M.E. Rothenberg, W.P. Jencks, *J. Am. Chem. Soc.*, 106, 1361 (1984).

[51] M.J. Kamlet, J.-L. M. Abboud, M.H. Abraham, R.W. Taft, *J. Org. Chem.*, 48, 2877 (1983).

[52] J.F. Bunnett, *J. Am. Chem. Soc.*, 83, 4956 (1961).

[53] J.P. Richard, W.P. Jencks, *J. Am. Chem. Soc.*, 106, 1373, 1383,1396 (1984).

[54] R.A. McClelland, F.L. Cozens, S. Steenken, T.L. Amyes and J.P. Richard, *J. Chem. Soc. Perkin Trans. 2*, 1717 (1993).

[55] V.B. Ivanov, V.L. Ivanov, M.G. Kuz'min, *Zh. Org. Khimii*, 8, 621 (1972) *(in Russian)*.

In: Reactions and Properties of Monomers and Polymers ISBN 1-60021-415-0
Editors: A. D'Amore and G. Zaikov, pp. 91-107

Chapter 5

REACTIONS OF OZONE WITH AROMATIC COMPOUNDS

S. D. Razumovskii
N.M. Emanuel's Institute of Biochemical Physics Russian Academy of Sciences
4, Kosygin str., Moscow 111999, Russia

INTRODUCTION

The cycle of investigations in the field of ozone chemistry in the Institute of Chemical Physics of USSR Academy of Sciences was started at the end of fifties of the XXth century in many respects due to N.M. Emanuel initiative. At the initial period the emphasis was placed on gas initiation of hydrocarbons oxidation processes, later on protection of rubber products from action of atmosphere ozone and only much later it became clear that reactions of ozone with hydrocarbons of various structures were fraught with a lot of outstanding possibilities.

Aromatic compounds are very important in person's life. From the one hand they are as useful companions accompany us everywhere: synthetic dyes, aspirin, plastics, fibers and many other products having aromatic cycles in their base. From the other hand, aromatic monsters generated by us such as polychlorinated aromatic hydrocarbons, dichlorodiphenyltrichloroethane (DDT) and dioxines turn to be strongest xenobiotics which are able to poison the environments in on the whole planet scale. Ozone from the one hand is often considered as effective mean of destruction of detrimental impurities of aromatic hydrocarbons in air (benzpyrenes) or water (phenols), from the other hand as powerful tool for functionalization of available product and infusion of new useful properties to it. The first mentions about ozone reactions with aromatic compounds relate to 1863 [1] when formation of CO_2 was observed under the action of ozone on benzoic acids in alkali medium. Later, in reaction of ozone with benzene the white solid sediment was obtained [2-9] formation of which was often accompanied by explosions. Similar sediments were also obtained in reactions of toluene with orto-xylene [8]. Garriers [7, 10, 11] have established the peroxide character of product and identified it as triozonide. Soon they found that benzene derivatives had the lowest reactivity in reactions with ozone in comparison with other classes of unsaturated hydrocarbons [12]. This caused the pointed discussion about structure of aromatic nucleus in homolytic raw of benzene and promoted developing of modern conceptions of aromaticity [11-15]. It was found in the framework of this discussion that under ozonation of orto-xylene the glyoxal, pyruvic aldehydes and diacetyl were formed [16]. This served as qualitative confirmation of C=C bonds delocalization in benzene cycle. Quantitative investigations showed [17-22] that independently on orto-xylene synthesis the products of its ozonolysis were always formed in ratio 3 : 2 : 1 which was expected under ozonolysis of equimolar mixture of hypothetic dimethylhexatrienes:

$$\text{(o-xylene, Kekulé form 1)} + \text{(o-xylene, Kekulé form 2)} + 6O_3 \xrightarrow{12[H]} 3\,OHC{-}CHO + CH_3COCOCH_3 + 2\,CH_3COCHO + 6\,H_2O$$

Examples of other investigations of ozone reactions with benzene derivatives are presented in works [23-27] and reviews [28-30].

In complex organic compounds aromatic cycles often present together with non-saturated fragments. When ozone influences on such compounds as a rule it attacks C=C bonds and only after their exhaustion reacts with aromatic nucleus [28, 31-40], rates of reactions reduce in the consecution: double bond in alkenes > triple bond in alkynes > benzene cycle. Such difference in reaction ability of various aromatic derivatives in relation to ozone creates large possibilities of synthesis of valuable products or destruction of harmful ones.

1. Complexes of Ozone with Benzene Homologues

At low temperatures ozone doesn't react with benzene and its homologues but forms π-complexes [41-43]. Nature of complexes, kinetics of their formation and consumption in reactions are investigated in [44-47]. Maximums of absorption of ozone π-complexes with a number of alkylbenzenes λ_{max} are situated in the diapason from 340 to 552 nm (Table 1) [43]. The following colours of π-complexes are observed: for benzene – reseda, anisole – olive-green, p-dineopentylbenzene – green, 1-ethyl-2-iodobenzene – green, isodurene – red-brown, 1,4-dimethylnaphthalene – purple. It is obvious from Table 1 that as potential of molecule ionization is reduced the value of λ_{max} is increased. Colour of π-complexes is also increased from green to purple s the length of absorption wave is increased.

In accordance with [43] absorption of π-complexes is caused by formation of complex with charge transfer in the system ozone (acceptor)–aromatic compound (π-donor). Complex has broad absorption spectrum in region 360-380nm whereas ozone and alkylbenzene in this region practically don't absorb. They showed in [44, 45] that ozone and alkylbenzene form complex of the structure 1 : 1:

$$ArX + O_3 \rightleftarrows ArX \cdot O_3$$

The data on some optic and other characteristics of complexes are presented in Table 1.

Table 1. Electron spectra of absorption of π-complexes of ozone with aromatic compounds. T = 78K, isopentane,

Compound	Ionization potential, (eV)	λ_{max}, nm	Colour
Benzene		340	Reseda
Toluene	8,82	383	Reseda
Ethylbenzene		382	Reseda
o-Xylene	8,56	410	
m-Xylene		390*	
Mesitine	8,40	441	
Tret-Butylbenzene		375	Reseda
Cumene		377*	
Pentamethylbenzene	7,92	541	Purple
Hexamethylbenzene		552	Purple

* – data from [45]; 216-263K, hexane,

Complex formation between ozone and substrate is reverse process. For example if the solution of ethylbenzene complex is heated from 78 up to 123K, then colour of solution is changed from green to blue. If solution of temperature is again reduced down to 78K green colour is restored. In other case the solution of pentamethylbenzene (which has the lowest ionization potential (Table 1)) was added to solution of ethylbenzene complex at 78K, this led to colour change from green to purple.

In Table 2 the data on stability of complexes in dependence on nature of substitution in benzene cycle are collected. As it is obvious from Table the value of equilibrium constant (K) is increased in the raw: chlorbenzene < fluorobenzene < benzyl chloride < benzene < ethylbenzene < toluene < cumene. Equilibrium constants are correlated with σ-m and σ-p constants of substituents according to Gammet [48]:

$$\lg K = \lg K_0 + \rho\sigma,$$

where K_0 – equilibrium constant of ozone complex with benzene, ρ – reaction constant. The correlation in the case of σ-m is fulfilled better than in the case of σ-p [47].

Kinetic and thermodynamic characteristics of complexes coordinate with the fact that ozone forms with benzene derivatives the complexes of structure 1 : 1. The determinant factor influencing on equilibrium constant $ArX + O_3 \rightleftarrows ArX{\cdot}O_3$ is inductive effect of substituents. π-complexes of ozone with ArX are both reversibly destructed into initial substances and may transform to various other products [47]. So, they assumed that destruction of ozone complex with cumene at 298-338K led t formation of ozonides, dimethylphenylcarbinole and free radicals.

It was shown [46] that the rate of complexes O_3-toluene and O_3-ethylbenzene consumption is described by kinetic equation:

$$W = -d[ArH \cdot O_3]_0 / d\tau = k_1[ArH \cdot O_3]_0 + k_2[ArH \cdot O_3]_0 [ArH]_0$$

The view of kinetic equation tells about the presence of two channels of complex consumption – destruction by first order and reaction with ArH. The values of $\mathbf{k_1}$ and $\mathbf{k_2}$ are presented in Table 3.

Table 2. Thermodynamic characteristics of complexes of O_3 with benzene and its derivatives (solvent is n-hexane) [45].

Compound	T, K	K, l/mole	ΔG^0*, kJ/mole	$-\Delta H^0$, kJ/mole	$-\Delta S$, J/mole·K
Cumene	253-259	1,00***	–	–	–
Toluene	263**	0,42**	1,89	9,6	44,4
	248	0,53			
	238	0,64			
	231	0,74			
Ethylbenzene	263	0,37	2,18	10,9	49,4
	247	0,46			
	238	0,62			
Benzene	263	0,21	3,44	–	–
Benzylchloride	263	0,15	4,15	–	–
Bromobenzene	263	0,10	5,03	5,90	41,5
	216	0,18	5,03	5,90	44,4
Fluorobenzene	263	0,07	5,82	5,90	44,4
Chlorobenzene	263	0,06	6,16	–	–

* – 263K; ** – the data were calculated from dependence of K on temperature; *** – [44]

Table 3. Constants of destruction of O_3 complexes with toluene and ethylbenzene at temperature 243K.

Compound	k_1, sec^{-1}	k_2, l/mole·sec
$C_6H_5CH_3$	$(1{,}78 \pm 0{,}29) \cdot 10^{-2}$	$(1{,}42 \pm 0{,}10) \cdot 10^{-2}$
$C_6H_5CH_2CH_3$	$(2{,}63 \pm 0{,}21) \cdot 10^{-2}$	$(1{,}25 \pm 0{,}11) \cdot 10^{-2}$

2. Mechanism of Ozone Reaction with Aromatic Nucleus

The products of ozone reaction with aromatic system of benzene cycle [10, 16, 18, 49-58] are usually ozonides of benzene or corresponding derivatives of other compounds of aromatic raw. Benzene triozonide – the product of addition of three ozone molecules to benzene is three-dimensional peroxide compound insoluble in known solvents and highly explosive. That is why researchers prefer to work with the system ozone–benzene in solutions with methanol additives when sediments are absent and the products of oxyperoxide character are formed instead of ozonides.

There is some uncertainty in identity of properties of ozonides of aromatic and aliphatic compounds that is confirmed also by the fact that name of ozonides includes the name of initial aromatic compound, for example benzene ozonide, mesitylene ozonide, although it is obvious that benzene and methylbenzene ozonation is accompanied by destruction of aromatic cycle. Formed ozonides should have mainly three-dimensional structure analogously to polymer ozonides formed in the case of cycloolefines [59]:

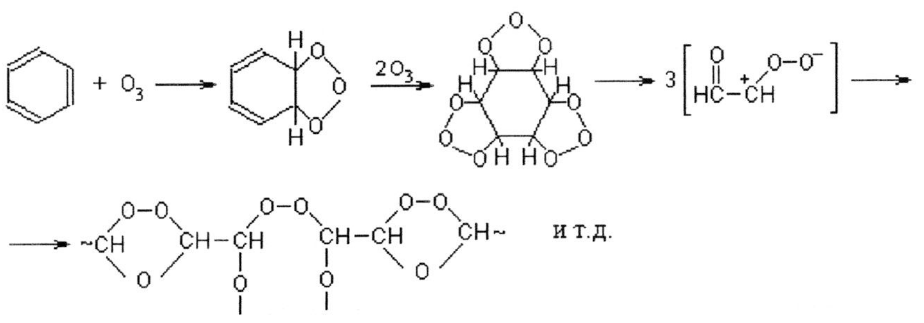

Among primary products of ozone reaction with benzene especially at high temperatures except ozonides and products of their destruction small amounts of phenol were obtained [60, 61]. They assumed that mechanism of phenol formation includes primary ozonide and epoxide [61]:

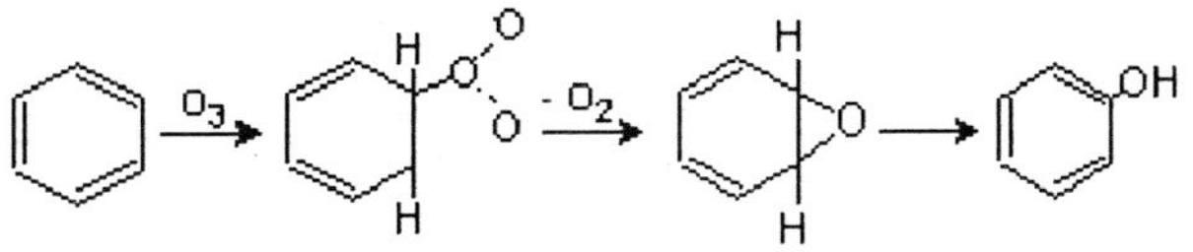

However in accordance with modern conceptions one considers more valid the mechanism which assumes formation of π-complex [43-46]. Detailed mechanism of ozone reaction with aromatic cycle leading to formation of phenol and other products may be presented by Scheme 1:

The Scheme includes π-complex formation as the first stage [30] and then this complex is transformed either into σ-complex (3), or gives ion-radical pair (2) which in the presence of active hydrogen is able to form radicals of $R^{\bullet}$ and $HO^{\bullet}$ type; σ-complex is transformed into phenol or molozonide (4) with formation of structure (6) able to regroup either into (7), or diarize into (11). Benzene triozonide (5) is formed by addition of two ozone molecules to molozonide (4). As a result of any of presented ways the polyozonide should be formed (12) – the main peroxide product of ozonolysis. Under benzene ozonation in solution of carbon tetrachloride instead of structure (12) the three-dimensional spatial structure (13) is more preferable [49]. Under ozonation in the medium of aliphatic acids [58, 62] in accordance with the mechanism of Krige [29, 30] the compound (7) may add acid with formation of acyloxyalkylhydroperoxide (14) which is polymerizationally inclined forming (15) [58].

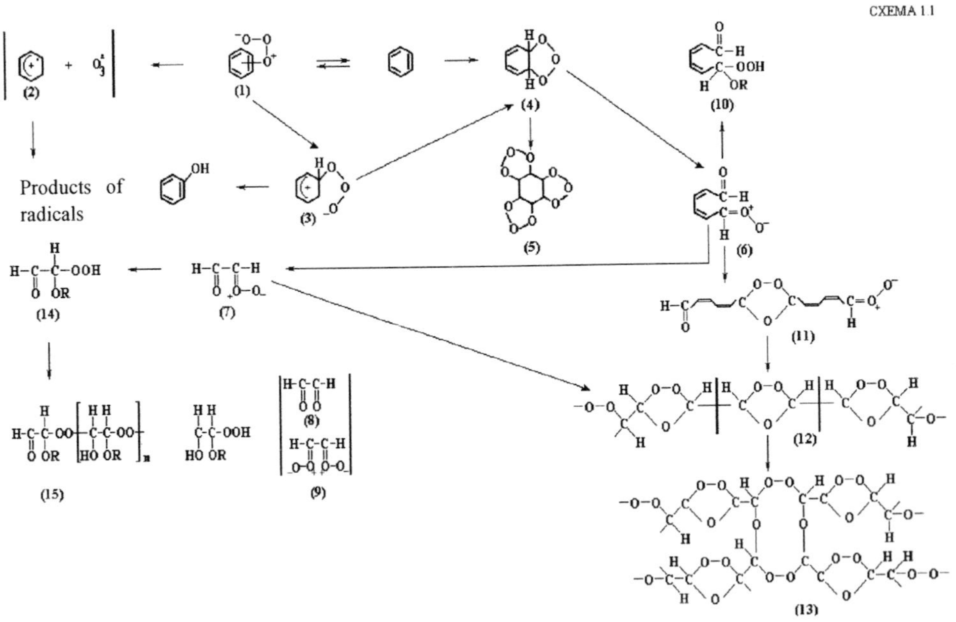

CXEMA 1.1

Products of radicals

Scheme 1.

3. Ozone Reactions with Alkylbenzenes without Involving of Aromatic Nucleus

As it was mentioned above the complex organic compounds under combined presence in molecule in spite of aromatic nucleus might contain also other groups, for example C=C, alkyl and others. If C=C bonds react as a rule as first ones, then aromatic and alkyl fragments turned to be comparable in their reaction abilities in relation to ozone [12]. Under ozone effect on such hydrocarbons the parallel, competitive with each other ozone reactions with aromatic cycle and side aliphatic chain proceed in system. Ratio of these two directions depends on alkyl benzene structure, reaction carryinf out conditions and ozidation depth [63-65]. So, in particular, toluene reacts with ozone in acetic acid. The main products of reaction are peroxide compounds. Benzyl alcohol, benzaldehyde and benzoic acid which are undoubtedly the products of methyl group oxidation are formed in fewer amounts (16%). Peroxides and benzoic acid are the final products of reaction. Benzyl alcohol and benzaldehyde are accumulated and consumed in the course of reaction. Peroxide compounds are stable enough to ozone action and are accumulated with high rate. Their yield per reacted toluene slightly depends on depth of substrate transformation and is equal to 80,3-84,6%.

Peroxides obtained after distillation under vacuum of acetic acid represent oily viscous liquid of straw colour, well soluble in acetic acid, but insoluble in dichlorethane and carbon tetrachloride. They energetically react with alkalis and potassium iodide. Infra-red spectra of extracted peroxides showed the absence of aromatic structures in their structure. Since mole ratio of reaction ozone–toluene is equal to 3 : 1, and judging by amount of active oxygen in peroxides we may assume that received peroxides are typical ozonides. Analogous results were obtained under ozonation of isomer xylenes 1,3,5– and 1,3,4-trimethylmenzene, 2,3,5,6-tetramethylbenzene [66, page 22].

Increase of a number of electron-donor substituents increases reaction ability of aromatic cycle. That is why tri- and tetramethylbenzenes react with ozone mainly by benzene cycle. So, under oxidation of 1,3,5-, 1,3,4-trimethylbenzenes and 2,3,5,6-tetramethylbenzene products of oxidation onserved aromatic nucleus were nor registered [66, page 23].

Under introduction into cycle of electron-acceptor substituents stability of aromatic cycle in reactions with ozone is increased.

At definite conditions ozone reaction with methylbenzene may be stopped at the stage of aromatic alcohols and aldehydes formation. One may be succeeded in this under ozonation of methylbenzene in the medium of acetic anhydride [67-69]. Analysis of literature data [70-78] leads to conclusion that primary reaction of ozone with cumene at low temperatures finally may also be presented by the scheme:

$CHMe_2$ + O_3 ⇄ $CHMe_2$ (O_3) → $CHMe_2$ (• +) + $O_3^{\bullet -}$

Electron transfer

H^+- Shear

$CHMe_2$ ← $CHMe_2$ (OOO^-)

$\overset{+}{C}Me_2$ + HO_3^-

Products of aromatic cycle ozonation

CMe_2OOOH

CMe_2OH + $COMe$ ← $CMe_2O^{\bullet}$ + $HO_2^{\bullet}$

In accordance with this scheme the π-complex was formed first which then was regrouped into products either of oxidation of alkyl group, or of addition to aromatic cycle.

Cumene hydrotrioxide formed at low temperatures further is destructed by radical-chain mechanism [70-71]:

$$ROOOH \rightarrow RO^{\bullet} + HO_2^{\bullet}$$

$$HO_2^{\bullet} + ROOOH \rightarrow H_2O_2 + ROOO^{\bullet}$$

$$ROOO^{\bullet} \rightarrow RO^{\bullet} + O_2,$$

which leads to cumene alcohol and acetophenone. Acetophenone obviously is received from β-phenylcumeneoxyl radicals:

$$RO^{\bullet} \rightarrow PhCOCH_3 + \dot{C}H_3$$

According with data of [79] the products ratio of cumene ozonation into cycle or into side chain at temperature 195-233K is equal to 30 : 70. So, it is evident that cumene ozonation rate into side chain is significantly higher than ozonation rate into cycle.

Formation of free radicals in ozone reaction with alkylbenzenes is observed directly while carrying out the reaction in cell of EPR-spectrometer [80]. It is additionally confirmed by intensive chemiluminescence appearing under ozone passing through alkylbenzenes [81]. Chemiluminescence is caused by interaction of peroxide radicals with each other.

4. Ozone Reactions with Polycyclic Aromatic Hydrocarbons

Close relations of benzene are polycyclic aromatic hydrocarbons (PAH): naphthalene, anthracene, phenanthrene and others and they are widely used in reception of dyes, lacquers, polymer compositions. So, it is not surprising that long ago chemists often use ozone for transformation of PAH into one or another valuable compound [82-84].

Reaction of ozone with PAH has a number of particularities. In particular, reaction with the second member of homogous raw naphthalene:

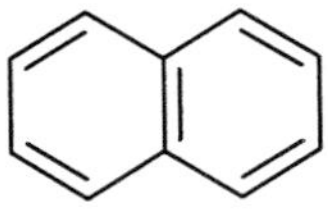

leads to formation of diozonide (1) in contrast to triozonide in the case of benzene ozonation (Scheme 2):

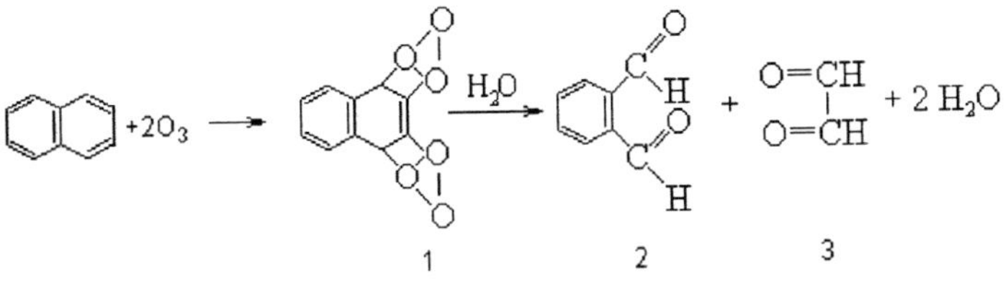

$+2O_3 \rightarrow$ 1 $\xrightarrow{H_2O}$ 2 + 3 $+ 2\,H_2O$

Scheme 2.

Diozonide under treatment by water gives o-phthaloaldehyde and glyoxal. Attack of naphthalene molecules by the first ozone molecule requires additional supplied energy for overcoming of conjugation of aromatic system. The second ozone molecule is added to in essence isolated C=C bond of defective cycle. Formed electro-negative fragments protect remaining aromatic nucleus from ozone action. Under carrying out of ozone reaction with naphthalene and its homologues in inert (aprotic) solvents the products of reaction (diozonide) form cross-linked structures which precipitate from solutions in the shape of white sediments. Their structure was the subject of several studies [28, 30, 85-88] results of which in integral view might be presented by the Scheme 3:

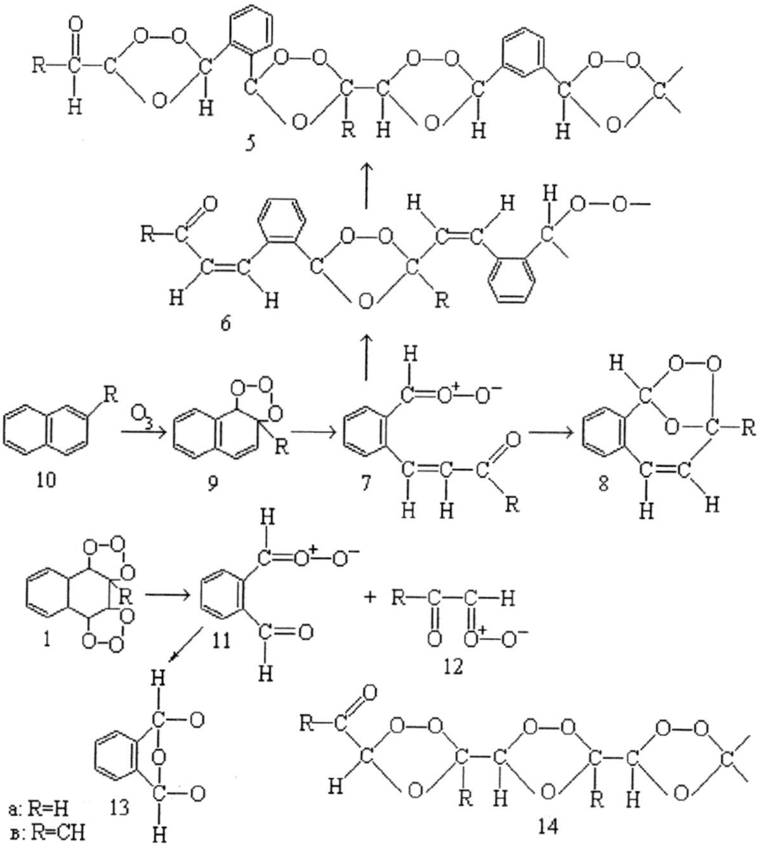

Scheme 3.

Reaction products are mainly consisted of mixture of non-regular co-polymers containing structures 16(5), 17(6), 18(8), 94 [66, pages 161, 87]. Some other products are formed under ozonation of naphthalene and substituted naphthalenes in proton-donor solvents [28, 30, 87]. Naphthalenes in methanol solvent as well as in aprotic solvents add two ozone molecules forming soluble peroxides which structure differs from ozonides (Scheme 4):

The most part of presented in Scheme 4 products are labile and are inclined to further transformations. In particular peroxide (22) was identified in reaction mixture at low temperatures. It is cyclizied with temperature rise with formation of 4-methoxy-2,3-benzdioxane-1-ol (24). While heating compounds (22) and (24) are transformed correspondingly into methyl ether of semi-aldehyde of phthalic acid and dimethyl phthalate. The last ones were identified by the method of gas-liquidchromatography [28, page 181]. In the mixture of products the following compounds were found: (22a) in amount of 3%, (22) – 70% and (23) – 7 mass %. This fact allows us concluding that induced influence of ozonide cycle formed at the previous reaction stage is displayed in preferable formation of bipolar ion at carbon atom most distant from the place of the first ozone molecule addition.

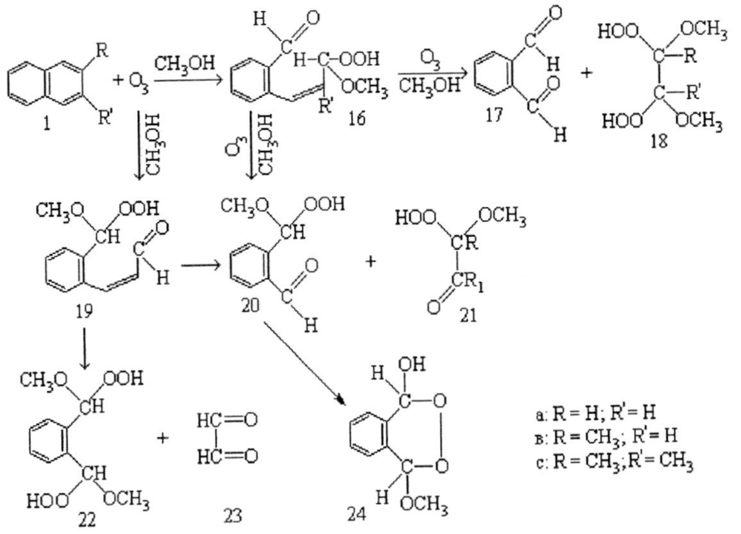

Scheme 4.

4-methoxy-2,3-benzdioxane-1-ol (24) may be relatively easy transformed into some very perspective derivatives. So, after distillation of ozonation products (24) is transformed into methylphthalaldehyde (27) (Scheme 5).

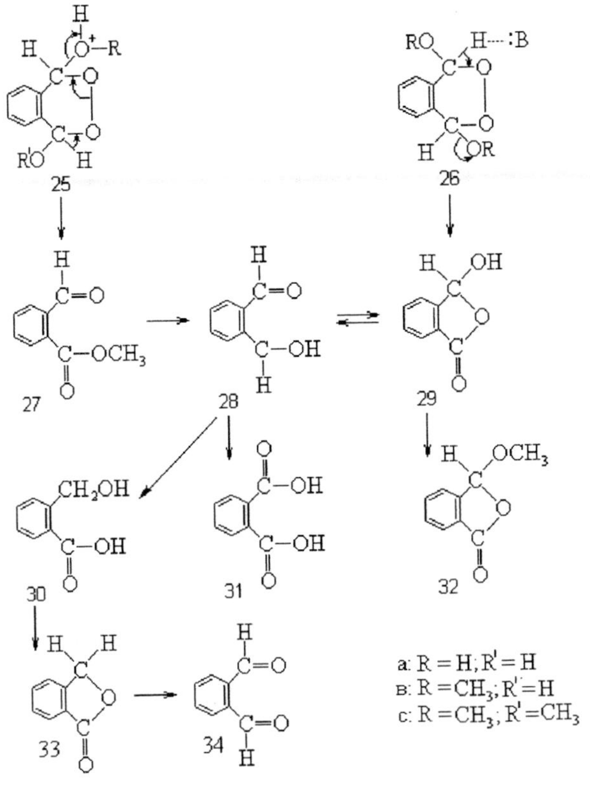

Scheme 5.

3-methoxyphthalaldehyde (32) may be isolated in the presence of hydrochloric acid, and after evaporation of reaction mixture and hydrochloric acid removal from suspensiuon the phthalaldehyde acid (28) was obtained with good yield. High yield of this acid is obtained also under reaction mixture treatment by water solution of sodium hydrochloride with further distillation and acidation. In the presence of hydrogen peroxide in acid or basic conditions the phthalic acid (31) is formed. In the presence of hydrogen peroxide decomposition of pure peroxide (24) leads to high yield of (28) and (31) [66, page 163].

The third member of the raw is phenanthrene

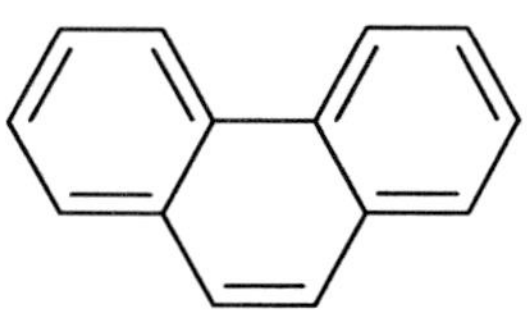

It reacts with ozone actively the same way as naphthalene forming polymer ozonide [28]:

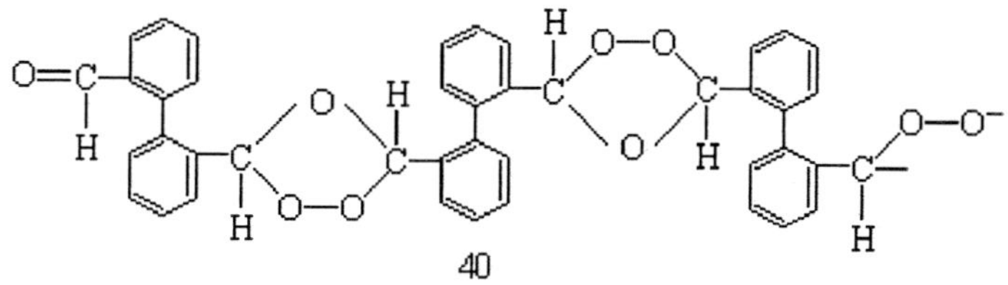

40

Particularities of structure of phenanthrene structure cause its ability to add one ozone molecule easily and after this reaction is strongly decelerated. As well as in the case of naphthalene this deceleration is caused by electro-negative induced effect of ozonide groups. In products of thermal and hydrolytic destruction 40 diphenic (41) and diphenaldehyde acids present in high yields.

Destruction 40 in the medium of strong oxidizing agent allows obtaining of 41 with high yield up to 96% [30, 94, 95].

The closest relation of phenanthrene is in essence the isomer anthracene:

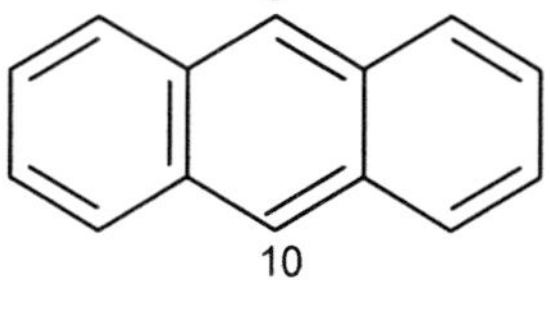

It reacts with ozone at positions 9 and 10 forming as the main product not ozonides, but anthracene [28, 30, 96]. In the course of reaction per one molecule of anthracene three ozone molecules are spent. Structure of products is complex and depends on reaction carrying out conditions especially on the nature of selected solvents [97, page 229].

The second product presenting in large amounts (18-67%) is phthalic acid. Anthrahydroquinone, semoquinone and 4,3-naphthalenedicarbonic acid which may be considered as a result of usual addition of ozone to C=C-bonds of aromatic cycle are also found in system in small amounts. It is also found that formation of anthraquinone is accompanied by molecular oxygen gassing, i.e. from ozone molecule the only one oxygen atom remains bonded in reaction products.

Formation of anthraquinone and anthracene initiated great interest to mechanism of this reaction and a number of works were published devoted to detailed investigation of products structure and mechanism of ozone reaction with anthrecene, the review of which was given in [66, page 168].

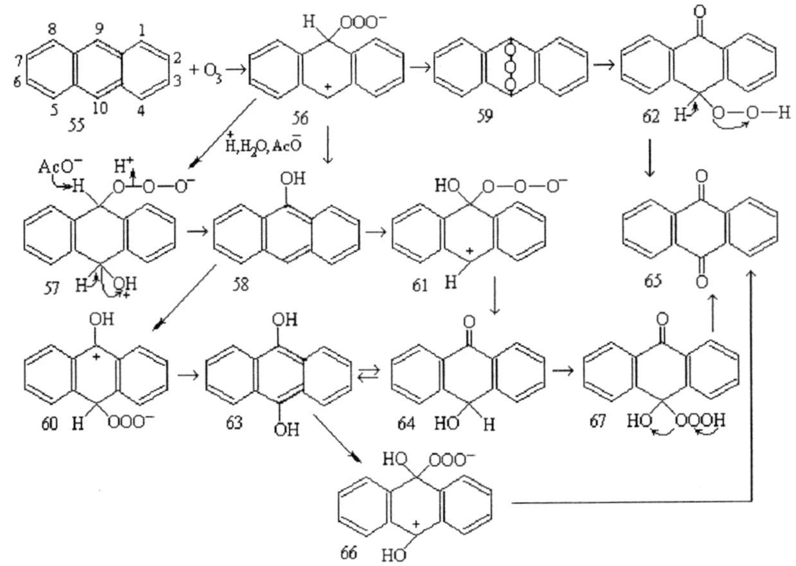

Scheme 6.

Direct attempts to observe free radicals by EPR method were not succeeded. However this fact doesn't exclude possibility of their presence in system since it is known that in process of low-molecular hydrocarbons oxidation the stationary concentration of free radicals as a rule is lower than limits of EPR spectrometer sensitivity.

Obviously, for final solution of the question whether the reaction is realized via the stage of hydrogen atom detachment or not we should compare ozone reaction with anthraquinone and ozone reaction with some anthraquinone analogue in which hydrogen in position 9 and 10 are substituted by atoms or groups not possessing the affinity to ozone for example by haloids atoms. It was shown on the example of ozone reaction that hydrogen substitution at C=C-bonds in olefins or aromatic systems by haloid doesn't hinder reaction proceeding, although reduces its rate [28, 98, 99]. It turned out under ozone action on 9,10-dibromide anthracene that reaction proceeds also with formation of anthraquinone [99]. 2,3-dibromidephthalic acid is formed parallel and its relative amount is higher than amount of 3,4-naphthalinedicarbonic acid under ozonation of anthracene. Result of this experiment allows excluding proposition about hydrogen removal from further consideration.

It was noticed above that in products of ozone reaction with the closest analogue of anthracene phenanthrene the phenanthrenequinone was presented. The scheme of its formation is also presented there.

If we permit that anthraquinone is formed analogously than basing on rate constants of ozone reaction with anthracene and intermediate products, for example carbonyl group ($5 \cdot 10^3$ and ~ 10 l/mole·sec accordingly) we should expect the appearance of anthraquinone after spending of more than 50% of anthracene. Both anthraquinone and carbonyl groups have characteristic bands in IR-spectra (1681 and 1705 cm^{-1} accordingly). This allows observing the kinetics of accumulation of both products by the change of intensity of mentioned bands in the course of ozone reaction with anthracene in carbon tetrachloride. In connection with the fact that anthraquinone was badly dissolved in CCl_4 the last one was removed from sample, sediment was dissolved in dioxane and the spectrum in the region 1600–1800cm^{-1} was registered.

It is obvious that anthraquinone is accumulated from primary stages of reaction. Accumulation of carbonyl products occurs in smaller amounts and is displayed in IR-spectra at later stages of process.

The last fact has good explanation if we assume that the first stage of reaction is electrophilic adition of ozone in positions 9 and 10. Further behavior of formed intermediate product (56) is caused by system properties and differs from cases of ozone addition to benzene or naphthalene described above by the fact that carbon atom situated near (C_{11} or C_{13}) enters the aromatic system which impacts it electro-negative properties in comparison with carbon atom in non-conjugated chain. Under consideration of particularities of ozone interaction with divisible bonds in which one of the atoms has high electro-negativity (bonds C=N, N=N or C=O) they noticed [100] that reaction proceeded with loss os two oxygen atoms, and sometimes with chain splitting:

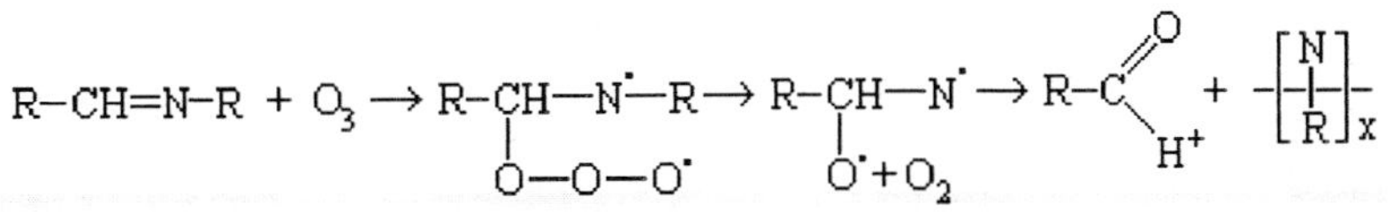

$$R\text{-}CH\text{=}N\text{-}R + O_3 \rightarrow R\text{-}\underset{O-O-O^{\bullet}}{\underset{|}{C}H}\text{—}\dot{N}\text{—}R \rightarrow R\text{-}\underset{O^{\bullet}+O_2}{\underset{|}{C}H}\text{—}N^{\bullet} \rightarrow R\text{-}C\overset{O}{\underset{H^+}{\lessgtr}} + \left[\underset{R}{\overset{N}{|}}\right]_x$$

Obviously some analogous phenomena occur also with intermediate reaction products (56) which then may transform in various ways (Scheme 7):

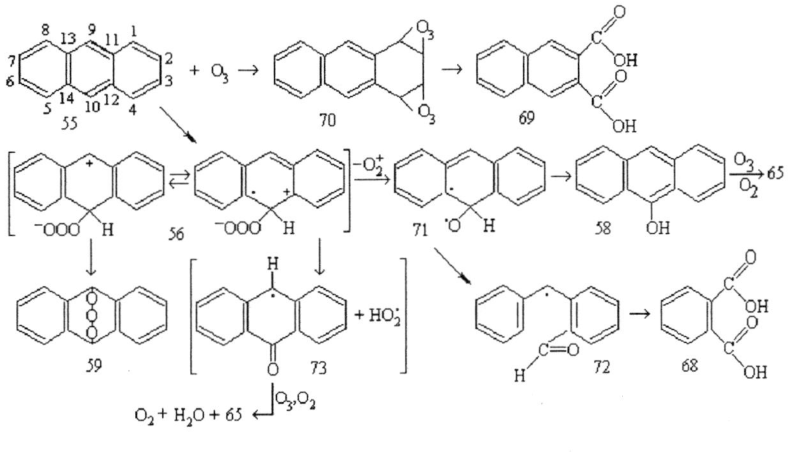

Scheme 7.

Shift of charge $+\delta$ or uncoupled electron to p-position may lead to formation of endoperoxide (29), and destruction of O–O-bond to (31), and hydrogen removal leads to formation of semiquinoide radical (30) which then may easily transform into anthraquinone (31) and one ozone molecule is spent at that. Kinetic particularities mentioned below (consumption of 2 moles of ozone per formation of 1 mole of anthraquinone, and the second molecule is added faster than the first) allow assuming that the main direction of process is consecutive transformation (25)–(26)–(30)–(31). The value of reaction rate constant equal to 43 l/mole·sec is obviously connected with ozone reaction with anthraquinone.

Thus, under investigation of ozone reaction with aromatic hydrocarbons two types of ozone addition to C=C bond of aromatic nucleus were found. In one case all three oxygens of ozone molecules are saved and the ozonides similar to olefins ozonides considered above are formed. In the other case in the molecule of novel compound only one oxygen atom of three is saved. Such type of addition is caused by the fact that one of the atoms at C=C-bond via which addition occurs has lower reactivity than other one due to one or other reasons. Together with addition reaction the reaction of hydrogen removal is observed in substituted aromatic hydrocarbons. This reaction is characteristic also for some other compounds – paraffin hydrocarbons, phenol, etc. [28].

Stated above data show how strongly insignificant differences in structure of various members of aromatic hydrocarbons family influence on reactions mechanisms. In the raw monocycle (benzene)–bicyclic (naphthalene)–tricyclic (phenanthrene) molecules in spite of general increase of reactivity the number of ozone molecules added at limiting stage of reaction is decreased in consecution 3 : 2 : 1. Two tricyclic isomers: phenanthrene and anthracene give as the main product of the first reaction act two various products: phenanthrene forms monoozonide, anthracene forms intermediate anthrol and anthron. Presence of substituents in aromatic cycles even greater broaden the set of primary and secondary products allowing together with molecular canals of transformation of labile intermediate product opening of free radical and chain ones. This significantly widened possibilities of ozone reaction with aromatic hydrocarbons application in both research practice and industry.

References

[1] E. Von Gorup-Besanez, *Liebigs Ann. Chem.*, 125, 207 (1863).

[2] A. Houreau, A. Renard, C.R., *Hebd Liebigs Ann. Chem.*, 170, 23 (1873); *Bull Soc. Chim. Fr.*, 19, No.2, 408 (1873).

[3] J.D. Boeke, *Ber. Dtsch. Chem. Ges.*, 6, 486 (1873).

[4] A.R. Leeds, *Ber. Dtsch. Chem. Ges.*, 14, 975 (1881).

[5] M. Mailfert, C.R., *Hebd Scances Acad. Sci.*, 94, 1186 (1882).

[6] L. Long, *Jr. Chem. Rev.*, 27, 437 (1940).

[7] C.D. Harries, *Liebigs Ann. Chem.*, 343, 311 (1905).

[8] A. Renard, C.R. Hebd, *Seances Acad. Sci.*, 120, 1177 (1895); ibid., 121, 651 (1895).

[9] M. Otto, *Ann. Chem. Phys.*, 13, No.7, 106, (1898).

[10] C.D. Harries, V. Weiss, *Ber. Dtsch. Chem. Ges.*, 37, 3431 (1904).

[11] C.D. Harries, V. Weiss, *Liebigs Ann. Chem.*, 343, 369 (1905).

[12] E. Molinari, *Ber. Dtsch. Chem. Ges.*, 40, 4154 (1907).
[13] C.D. Harries, *Ber. Dtsch. Chem. Ges.*, 40, 4905 (1907).
[14] E. Molinari, *Ber. Dtsch. Chem. Ges.*, 41, 585, 2782 (1908).
[15] C.D. Harries, *Ber. Dtsch. Chem. Ges.*, 41, 1227 (1908).
[16] A.A. Levine, A.G. Cole, *J. Am. Chem. Soc.*, 54, 338 (1932).
[17] J.P. Wibaut, P.W. Haaijman, *Nature (London)*, 144, 290 (1939); *Science (Washington O.C.)*, 94, 49 (1941).
[18] P.W. Haaijman, J.P. Wibaut, *Recl. Trav. Chim. Pays-Bas.*, 60, 842, (1941).
[19] J.P. Wibaut, *Bull. Soc. Chim. Fr.*, 996 (1950).
[20] J.P. Wibaut, *J. Chim. Phys. Biol.*, 53, 111 (1956).
[21] J.P. Wibaut, *End. Chim. Belge*, 20, 3 (1955).
[22] J.P. Wibaut, *Chimia*, 11, 298 (1957).
[23] P.S. Bailey, *Chem. Rev.*, 58, 925 (1958).
[24] E. Fonrobert, *Das. Ozon. Enke. Stuttgart*, (1916).
[25] E. Fonrobert, *in Die Methoden der Organischen chemie (J. Houben. ed.) 3 rd.*, 3, 406, Thieme, Leipzig (1930).
[26] R. Wilstätter, *Ber. Dtsch. Chem. Ges.*, 59, 123 (1926).
[27] E.C. Kooyman, J.A.A. Ketelaar, *Recl. Trav. Chim., Pays-Bas*, 65, 859 (1946).
[28] S.D. Razumovskii, G.E. Zaikov, *Ozone and its reactions with organic compounds*, Moscow: Nauka (1974) *(in Russian)*.
[29] S.D. Razumovskii, D.M. Shopov, S.K. Rakovskii, G.E. Zaikov, *Ozone and its reactions with organic compounds*, Sophia, Bulgaria AN, 287 (1983).
[30] P.S. Bailey, *Ozonation in organic chemistry*, 2, *Nonolefinic Compounds*, N.-Y., L., Academic Press., 497 (1982).
[31] Pat. USA No.3.958.942 (1976).
[32] C.D. Harries, R. Haarmann, *Ber. Dtsch. Chem. Ges.*, 98, 32 (1915).
[33] C.D. Harries, A.S. de Osa, *Ber. Dtsch. Chem. Ges.*, 37, 842 (1904).
[34] P.S. Bailey, *Chem. Ber.*, 87, 993 (1954).
[35] J.L. Warnell, R.L. Shriner, *J. Am. Chem. Soc.*, 79, 3165 (1957).
[36] P.J. Garratt, K.P.C. Vollhardt, *Synthesis*, 423 (1971).
[37] R. Criegee, P. de Pruyn, G. Lohaus, *Liebigs Ann. Chem.*, 583, 19 (1953).
[38] G. Brus, G. Peyresblanques, C.R. Hebd, *Seances Acad. Sci.*, 190, 685 (1930).
[39] A.I. Yakubchik, N.G. Kasatkina, T.E. Pavlovskaya, *Zh. Obshei Khimii*, 25, 1473 (1955) *(in Russian)*.
[40] S.D. Razumovskii, G.E. Zaikov, *Zh. Org. Khimii*, 8, 468 (1972).
[41] P.S. Bailey, J.W. Ward, *J. Am. Chem. Soc.*, 93, 3552 (1971).
[42] L.A. Hull, I.C. Hisatsune, J. Heicklen, *J. Am. Chem. Soc.*, 94, 4856 (1972).
[43] P.S. Bailey, J.W. Ward, T.P. Carter, Jr.E. Nich, C.M. Ficher, A–I.V. Khashab, *J. Am. Chem. Soc.*, 96, 6136 (1974).
[44] V.V. Shereshovetz, V.D. Komissarov, L.G. Galimova, *Izv. AN SSSR, Ser. Khim.*, 2632 (1980) *(in Russian)*.
[45] V.V. Shereshovetz, L.G. Galimova, V.D. Komissarov, *Izv. AN SSSR, Ser. Khim.*, 2488 (1981) *(in Russian)*.
[46] R.K. Yanbaeva, V.V. Shereshovetz, V.D. Komissarov, *The IId All-union conference on ozone synthesis and application*, Moscow, 87 (1991) *(in Russian)*.
[47] L.J. Andrews, R.M. Keefer, *J. Am. Chem. Soc.*, 72, 3113 (1950).

[48] L. Gammlet, *The basis of physical organic chemistry*, Moscow: Mir (1972) *(in Russian)*.

[49] S.D. Razumovskii, G.E. Zaikov, *Izv. AN SSSR, Ser. Khim.*, No.12, 2657 (1972) *(in Russian)*.

[50] P.S. Bailey, S. Bath, F. Dobinson, F. Garcia-Scharp, C.D. Jonson, *J. Org. Chem.*, 29, 697, 703 (1964).

[51] M.G. Sturrock, B.J.Cravy, V.A. Wing, *Canad. J. Chem.*, 49, 3047 (1971).

[52] P.S. Bailey, B.M. Mainthia, *J. Org. Chem.*, 23, 1089 (1958).

[53] P.S. Bailey, *J. Org. Chem.*, 22, 89 (1957); 29, 1400, 1409 (1964).

[54] P.S. Bailey, J.E. Batterbee, A.G. Lange, *J. Am. Chem. Soc.*, 90, 1027 (1969).

[55] Noller C.R., *J. Am. Chem. Soc.*, 57, 2442 (1935).

[56] N.D. Rus'yanova, L.P. Yurkina, N.S. Popova, *Koks i khimiya*, No.5, 41 (1969) *(in Russian)*.

[57] H. Rupe, H. Hirschmann, *Helv. chim. Acta.*, 14, 49 (1931).

[58] T.W. Nakagava, L.J. Andrews, R.M. Keefer, *J. Am. Chem. Soc.*, 82, 269 (1960).

[59] S.D. Razumovskii, I.A. Titorskii, Yu.N. Yur'ev, G.A. Niazashvilly, *Vysokomol. Soed.*, 13A, 197 (1971) *(in Russian)*.

[60] M. Nencki, P. Giacosa, Z. Hoppe Seyler's, *Physiol. Chem.*, 4, 339 (1880).

[61] V.D. Komissarov, I.N. Komissarova, *Izv. AN SSSR, Ser. Khim.*, 677 (1973) *(in Russian)*.

[62] E. Bernatek, E. Karlsen, T. Ledaal, *Acta Chem. Scand.*, 21, No.5, 1229 (1967).

[63] G.A. Galstyan, T.M. Galstyan, S.M. Sokolova, *Kinetika i kataliz*, 33, No.4, 779 (1992) *(in Russian)*.

[64] G.A. Galstyan, *Diss. Of Dcotor of chemical sciences*, Lvov (1992) *(in Russian)*.

[65] G.A. Galstyan, *Zh. Fizich. Khimii*, 66, No.4, 875 (1992) *(in Russian)*.

[66] G.A. Galstyan, N.F. Tyupalo, S.D. Razumovskii, *Ozone and its reactions with aromatuic compounds in liquid phase*, Lugansk (2004) *(in Russian)*.

[67] G.A. Galstyan, L.A. Matzegora, O.I. Rister, *Zh. Prikl. Khimii*, 55, No.11, 2547 (1982) *(in Russian)*.

[68] G.A. Galstyan, V.V. Vedernikov, A.N. Gudym, *Khimiya i khimicheskaya technologiya*, 26, No.9, 1041 (1983) *(in Russian)*.

[69] E.V. Potapenko, G.A. Galstyan, *Ukr. Khim. Zh.*, 66, No.1, 34 (2000) *(in Russian)*.

[70] W.A. Pryor, G.J. G leicher, D.F. Church, *J. Org. Chem.*, 49, No.14, 2574 (1984).

[71] P.S. Nangia, S.W. Benson, *J. Am. Chem. Soc.*, 102, No.9, 3105 (1980).

[72] V.D. Komissarov, *Diss. Of Dcotor of chemical sciences*, Ufa (1990) *(in Russian)*.

[73] V.V. Shereshovetz, N.Ya. Shafikov, V.D. Komissarov, *Zh. Fizich. Khimii*, 54, No.5, 1288 (1980) *(in Russian)*.

[74] V.V. Shereshovetz, N.Ya. Shafimov, V.D. Komissarov, *Kinetika i kataliz*, 21, No.6, 1596 (1980) *(in Russian)*.

[75] V.V. Shereshovetz, N.Ya. Shafikov, G.S. Lomakin, *Izv. AN SSSR, Ser. Khim.*, No.6, 1265 (1985) *(in Russian)*.

[76] W.A. Pryor, D. Giamalva, D.F. Church, *J. Am. Chem. Soc.*, 107, No.9, 2793 (1985).

[77] D. Giamalva, D.F. Church, W.A. Pryor, *Biochem and Biophys, Res. Commun*, 133, No.2, 773 (1985).

[78] E.V. Avzyanova, N.N. Kabal'nova, V.V. Shereshobetz, *Izv. RAN, Ser. Khim.*, No.2, 371 (1996) *(in Russian)*.

[79] W.A. Pryor, G.J. Gleicher, D.F. Church, *J. Org. Chem.*, 48, 3614 (1983).
[80] S.K. Rakovskii, S.D. Razumovskii, G.E. Zaikov, *Izv. AN SSSR, Ser. Khim.*, No.3, 701 (1976) *(in Russian).*
[81] V.Ya. Shlyapintokh, A.A. Kefeli, V.I. Gol'denberg, S.D. Razumovskii, *Dokl. AN SSSR*, 186, 1132 (1969) *(in Russian).*
[82] S.D. Razumovskii, Inventors certificate of USSR No.235759.
[83] I.L. Boguslavskaya, L.I. Kozhushkova, L.A. Kozorez, V.A. Yakobi, Inventors certificate of USSR No.319212.
[84] S.D. Razumovskii, L.V. Berezova, Yu.N. Yur'ev, Inventors certificate of USSR No. 240700.
[85] C.D. Harries, V. Weiss, *Liebigs Ann. Chem.*, 343, 369 (1905).
[86] L. Long, *Jr. Chem. Rev.*, 27, 437 (1940).
[87] P.S. Bailey, S.S. Bash, F. Dobanson, F.J. Garsia-Sharp, C.D. Johnson, *J. Org. Chem.*, 29, 697 (1964).
[88] P.S. Bailey, Chem. Rev., 58, 925 (1958).
[89] C.D. Harries, V. Weiss, *Liebigs Ann. Chem.*, 343, 369 (1905).
[90] L. Long, *Jr. Chem. Rev.*, 27, 437 (1940).
[91] P.S. Bailey, *Chem. Rev.*, 58, 925 (1958).
[92] R.W. Murray, R.D. Voussefych, P. Story, *J. Am. Chem. Soc.*, 89, 2429 (1967).
[93] S.D. Razumovskii, Yu.N. Yur'ev, *Neftekhimiya*, 6, 737 (1966) *(in Russian).*
[94] N.D. Rus'yanova, V.G. Koksharov, *All-bion Conference on catalytic reactions in liquid phase*, Alma-ata, 433 (1963) *(in Russian).*
[95] V.T. Kalakutzkii, N.D. Rus'yanova, *Khim. Prod. Koks. Uglei*, No.5, 232 (1969) *(in Russian).*
[96] I.M. Roitt, W.A. Waters, *J. Chem. Soc.*, 3060, (1949).
[97] S.D. Razumovskii, G.E. Zaikov, *Ozone and its Reactions with Organic Compounds*, Elsevier, Amsterdam (1984).
[98] S.D. Razumovskii, L.M. Reutova, G.A. Niazashvilly, I.A. Tutorskii, G.E. Zaikov, *Dokl. AN SSSR*, 194, 1127 (1970) *(in Russian).*
[99] P. Kolsaker, P.S. Bailey, F. Dobinson, B. Kumar, *J. Org. Chem.*, 29, 1409 (1964).
[100] S.D. Razumovskii, G.E. Zaikov, *Zh. Org. Khimii*, 8, 468 (1972) *(in Russian).*

In: Reactions and Properties of Monomers and Polymers
Editors: A. D'Amore and G. Zaikov, pp. 109-123
ISBN 1-60021-415-0

Chapter 6

APPLICATION OF A MODEL BASED ON CONSECUTIVE REACTIONS TO POLYMER DEGRADATION

Alfonso Jiménez[a], Rafael Balart[b], Nuria López[c] and Juan López[b,*]

[a] Department of Analytical Chemistry, Nutrition and Food Sciences. University of Alicante, PO Box 99, 03080, Alicante, Spain

[b] Department of Mechanical and Materials Engineering, Polytechnic University of Valencia, Paseo del Viaducto 1, 03801 Alcoy, Alicante, Spain

[c] Technological Institute of Plastics (AIMPLAS), Technological Park of Valencia, Gustave Eiffel, 4, 46980, Valencia, Spain

ABSTRACT

Thermal degradation models are very useful in the prediction of the polymer response under certain temperature conditions. Traditional kinetic models are being questioned, since they show some drawbacks when applied to simple degradation processes. Considering the basic expression for polymer degradation processes, we have developed a new model and it has been validated by comparison with traditional kinetic methods. The proposed model is based on the consideration of polymer degradation processes as described by two consecutive reactions. The resolution of the proposed equations by this method is simple and gives a dependency of the conversion degree as a function of time and temperature which fits quite accurately a sigmoidal plot in a wide range of experimental data. A very good match between experimental results obtained by thermogravimetric analysis of poly(acrylonitrile-butadiene-styrene), ABS, and polycarbonate, PC, and data obtained by application of the proposed model was observed.

Keywords: Thermogravimetric analysis; polymer degradation; apparent activation energy; consecutive reaction model; kinetic analysis

* Corresponding autor. Tel./Fax: +34 96652 84 78; E-mail address: jlopezm@mcm.upv.es (J. López)

1. INTRODUCTION

The study of thermal degradation mechanisms of polymers allows us to predict the lifetime of these materials in certain temperature conditions [1]. Polymers normally suffer some transformation processes that imply the heating of the material to a certain temperature, near degradation in some cases. This can be the origin of certain degradation of the polymer and the cause of a partial modification of the material properties before use. There is a wide range of processing temperatures in some polymers before degradation occurs, but in other cases this range is clearly narrowed for high temperatures. In this case it is necessary the use of thermal stabilizers and a precise knowledge of the degradation mechanism will allow us to select the most suitable stabilizing system and processing conditions.

The first kinetic models to describe degradation processes in polymers were proposed during the sixties [2,3]. These models have been widely used up to the present [4-6], although their limitations have been reported and some alternatives have been recently proposed [7]. The classical methods are based on the basic rate equation that represents the conversion rate, $d\alpha/dt$, at fixed temperatures as a function of the conversion degree (α) (equation 1).

$$\frac{d\alpha}{dt} = K \cdot f(\alpha) \tag{1}$$

In polymer degradation it is supposed that the conversion degree is proportional to the concentration of the non-reacted polymer, and the combination with the Arrhenius expression leads to the fundamental equation of methods to calculate the kinetic parameters from thermogravimetric data (equation 2).

$$\frac{d\alpha}{(1-\alpha)^n} = \frac{A}{\beta} \cdot e^{\left(\frac{-Ea}{R \cdot T}\right)} \cdot dT \tag{2}$$

where β is the heating rate, A the preexponential factor and Ea the apparent activation energy. There is another simple expression of the polymeric degradation kinetics, which is known as autocatalytic model, where α is related to kinetic parameters, n and m, as shown in equation 3.

$$f(\alpha) = (1-\alpha)^n \cdot \alpha^m \tag{3}$$

where n and m are constants indicative of two complementary reaction orders. The general expression of the degradation kinetics can be considered as in equation 4.

$$\frac{d\alpha}{dT} = \frac{A}{\beta} \cdot e^{\left(\frac{-Ea}{R \cdot T}\right)} \cdot (1-\alpha)^n \cdot \alpha^m \tag{4}$$

The application of this model gave good results for the description of the polymeric degradation behaviour in some cases [8,9]. However, its indiscriminate use is not recommended as it shows some problems:

- It is not applicable to the whole degradation range in many polymers.
- The physical meaning of "n" and “m” is not clear in some cases.

Criado and Perez-Maqueda [10] reported that different solutions show the same validity for the interpretation of experimental results. Consequently, the information supplied by the application of different kinetic models has a limited physical significance, as the autocatalytic model can be considered just an empirical interpretation of results. Therefore it is necessary to reconsider the use of the classical kinetic models of thermal degradation in some cases in order to obtain reliable information of the process. Good fittings can only be obtained with a previous knowledge of the mechanism of reaction [11-13] and new models, like the "multiple reaction model" (MRM) [14] or those based on artificial neural networks [15] have been proposed.

2. Theoretical Background

One of the main shortcomings of the thermogravimetric analysis applied to the determination of kinetic parameters by using classical models is that volatiles evolved from the degradation processes are only a fraction of those corresponding to the chemical mechanism of the degradation of a polymeric material. Therefore, it is necessary to consider other reactions to get the complete vision of those processes occurring at high temperatures.

The simplest hypothesis to solve this problem is to assimilate the degradation process to several consecutive reactions. When we follow the evolution of a decomposition product (W) from a polymer (P), we can suppose that at the first stage both chemical species are in direct contact inside the reaction chamber and close to the polymer, which decomposes at a speed characterized by a constant (K_1). The evolved compound(s) remain in the interior of the reaction chamber and may form a gas (W_3), which is able to evolve from the polymer with a speed described by a constant (K_2). This loss of weight is determined experimentally in the thermogravimetric balance. A scheme as described in equation (5) can be supposed

$$P_1 - W_1 \xrightarrow{K_1} P_1 * W_2 \xrightarrow{K_2} P_1 + W_3 \uparrow \tag{5}$$

Consequently by application of kinetic equations we can calculate the amount of each compound in the course of the decomposition reaction.

$$\begin{aligned} & W_1 + W_2 + W_3 = W_0 \\ & \frac{dW_1}{dt} + \frac{dW_2}{dt} + \frac{dW_3}{dt} = 0 \\ & \frac{dW_1}{dt} = -K_1 W_1 \end{aligned} \tag{6}$$

$$\frac{dW_2}{dt} = K_1 W_1 - K_2 W_2$$

$$\frac{dW_3}{dt} = - K_2 W_2$$

By substitution we can calculate the amount of each compound evolved from the polymer:

$$W_1 = W_0 e^{-K_1 \Delta t}$$
$$W_2 = W_0 \left(\frac{K_1}{K_2 - K_1} \right) \left[e^{-K_1 \Delta t} - e^{-K_2 \Delta t} \right] \tag{7}$$
$$W_3 = W_0 \left[1 - \frac{1}{(K_2 - K_1)} \left(K_1 e^{-K_2 \Delta t} - K_2 e^{-K_1 \Delta t} \right) \right]$$

If we only focus our attention on the final product, we can join all the equations in the basis of the calculation of the conversion degree:

$$\alpha = \left[1 - \frac{1}{(K_2 - K_1)} \left(K_1 e^{-K_2 \Delta t} - K_2 e^{-K_1 \Delta t} \right) \right] \tag{8}$$

This sigmoidal equation depends on two parameters (K_1 and K_2). Although the results of this equation are symmetrical for low conversion degrees, both parameters are interchangeable. The form of the curve depends on two factors: the values of each constant, characteristic of each reaction, and on the other hand the relation between them (Fig 1 and 2).

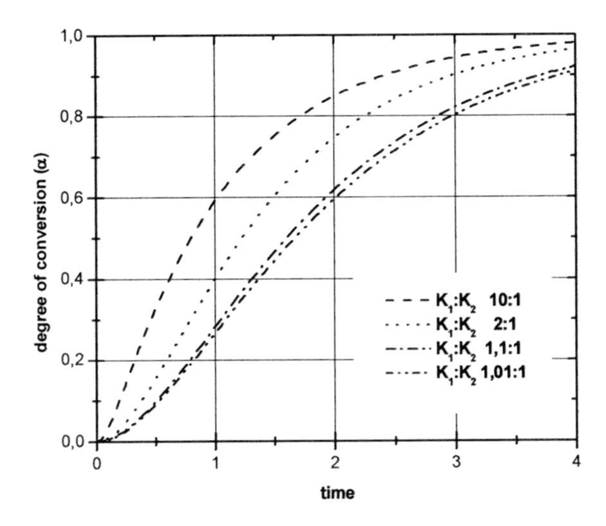

Figure 1. Evolution of the curves for different values of the constants, keeping one of them in a constant value to 1.

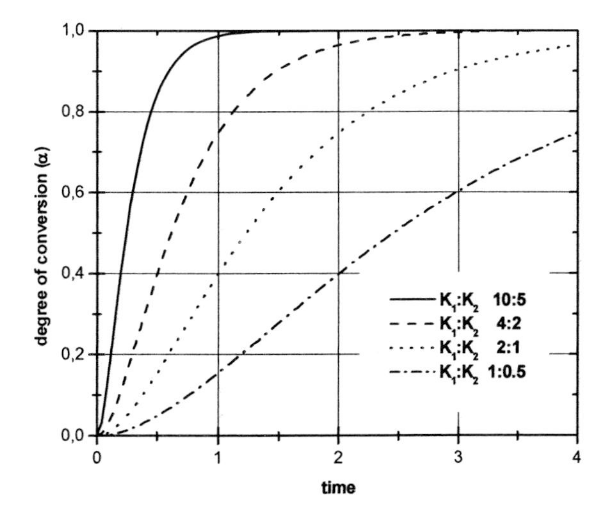

Figure 2. Evolution of the curves for different values of the constants, keeping constant the ratio between them to 2:1.

We can consider the two limiting situations. Firstly, we can consider the case that the difference between both constants is very large. In this case the equation can be expressed on the following way (equation 9):

$$\alpha = \left[\frac{K_1\left(1-e^{-K_2 \Delta t}\right) - K_2\left(1-e^{-K_1 \Delta t}\right)}{(K_1 - K_2)} \right] \quad (9)$$

$$\text{if } K_1 >> K_2 \;\Rightarrow\; \alpha \cong \left(1-e^{-K_2 \Delta t}\right)$$

Finally an exponential curve corresponding to a one-step degradation model is obtained:

$$\frac{d\alpha}{dt} = K(T)[1-\alpha] \quad (10)$$

On the other hand, if both constants are similar, we can change variables by defining the difference between constants as δ.

$$K_1 - K_2 = \delta$$

$$\alpha = \left[1 - \frac{1}{\delta}\left[K_1 e^{-(K_1-\delta)t} - (K_1-\delta)e^{-K_1 t}\right]\right]$$

$$\alpha = \left[1 - \frac{K_1 e^{-(K_1-\delta)t} - (K_1-\delta)e^{-K_1 t}}{\delta}\right] \quad (11)$$

$$\alpha = \left[1 - \frac{K_1 e^{-K_1 t} e^{\delta t} - K_1 e^{-K_1 t} + \delta e^{-K_1 t}}{\delta}\right]$$

$$\alpha = \left[1 - e^{-K_1 t}\left(\frac{K_1 e^{\delta t} - K_1 + \delta}{\delta}\right)\right]$$

We can calculate the limit of the expression when δ tends to zero

$$\lim_{\delta \to 0} F(\delta) = \frac{K_1 e^{\delta t} - K_1 + \delta}{\delta} \to 1 + K_1 t \tag{12}$$

Finally, the equation results in:

$$\alpha = 1 - e^{-K_1 \Delta t}\left(1 + K_1 \Delta t\right) \tag{13}$$

If we plot these equations we also obtain sigmoidal curves, whose form depends on the K value (Figure 3). We can see from the previous results that equation (13) can be applied to the results obtained by isothermal thermogravimetry. The following step is the study of the applicability of this model to dynamic thermogravimetry, where the temperature is not constant during the test. In a usual program, temperature increases with time, with a linear slope defined by the heating rate, β

$$T = T_0 + \beta t \tag{14}$$

This dependency between temperature and time difficults to obtain the general expression from the basic differential equation. The resolution of such equations makes necessary the use of some approximations only applicable to a part of the process, not to the whole like the classic models.

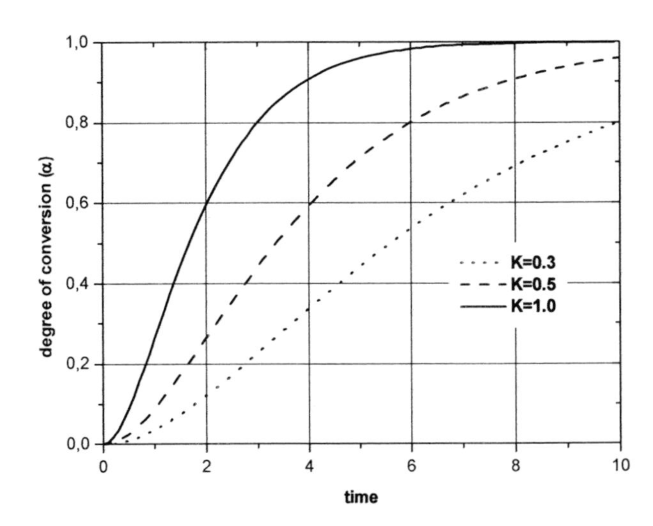

Figure 3. Evolution of the curves for different values of the constants.

MacCallum and Tanner considered that the conversion degree is a function of temperature and time [16]

$$\alpha = f(T,t) \tag{15}$$

And the differential equation can be converted in the following expression:

$$\left(\frac{d\alpha}{dt}\right) = \left(\frac{\partial \alpha}{\partial t}\right)_T + \left(\frac{\partial \alpha}{\partial T}\right)_t \left(\frac{\partial T}{\partial t}\right)_\alpha \tag{16}$$

Some authors proposed a solution to this equation [17-19], as they considered that there is no change in α with time. Therefore the last part of the second term in equation (16) is equal zero, and it is only necessary to solve the following expression:

$$\left(\frac{d\alpha}{dt}\right) = \left(\frac{\partial \alpha}{\partial t}\right)_T \tag{17}$$

The final result is equation (10), where K shows temperature dependency

$$\alpha = \left[1 - \frac{1}{[K_1(T) - K_2(T)]}\left(K_1(T)e^{-K_2(T)\Delta t} - K_2(T)e^{-K_1(T)\Delta t}\right)\right] \tag{18}$$

Nevertheless, it is very difficult to obtain valuable information from this equation if considering the variability of the two constants. Thus, it is necessary to propose another approximation. If temperature increases, the rate of the slowest process is the determining step and consequently we can consider a process where the constants tend to equal themselves according to equation (15).

$$\alpha = 1 - e^{-K(T)_1\ \Delta t}\left(1 + K(T)\Delta t\right) \tag{19}$$

By introduction of the Arrhenius exponential expression

$$K = Ae^{\left(\frac{-Ea}{RT}\right)} \tag{20}$$

We can plot the theoretical curves of the conversion degree of the reaction as a function of ln A, Ea and β at different temperatures:

$$\alpha(T,\beta) = 1 - \left[e^{-K(A,E,T)\frac{T}{\beta}}\right]\left[1 + K(A,E,T)\frac{T}{\beta}\right] \tag{21}$$

It can be observed that equation (21) reproduces the experimental shift of the curves with the heating rate, which is more important for the lowest rates (Figure 4).

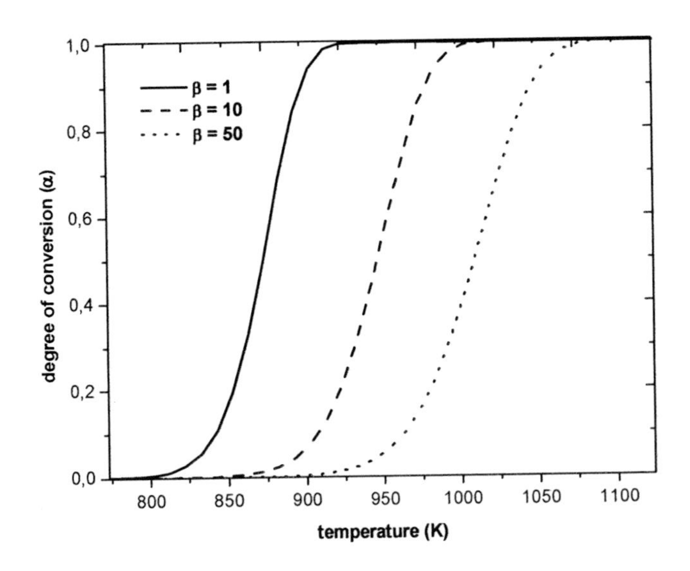

Figure 4. Plots of α(T) at different values of heating rate β (1, 10, 50 K/min). Model parameters: LnA = 14, E_a = 100 kJ/mol.

3. Experimental

Some experiments were run for ABS and Polycarbonate (PC) provided by ACTECO S.A (Alcoy, Spain). Both materials were characterized and the obtained results for MFI, density and VST were as follows. For polycarbonate: MFI (230 °C/5kg) = 6,06; density = 1,2 g/cm^3; VST = 139,2°C. For ABS: MFI (230°C/5kg) = 20,45; density = 1,05 g/cm^3; VST = 95,3°C.

Thermal tests were carried out by using a thermogravimetric balance Mettler-Toledo 851e-TGA-SDTA (Mettler Toledo Inc., Schwarzenbach, Switzerland). Dynamic tests at several heating rates (2, 5, 10, 15, 20, 25 and 30 °C/min) were carried out to determine weight loss.

4. Results and Discussion

In order to validate the proposed model we have applied it to two industrially important materials like PC and ABS. This selection was based on their different thermal degradation behaviour. ABS shows a degradation mechanism based on constant kinetic parameters of decomposition for the whole range of heating rates, but in the case of PC we observed that the kinetic parameters depend on the heating rate. This can be observed in the degradation curves, as the variation with the rate of the test is very symmetrical for the ABS (Figure 5) and greater asymmetry for the polycarbonate (Figure 6).

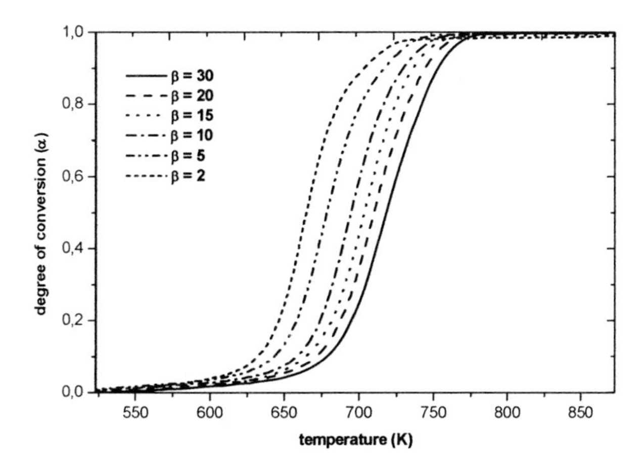

Figure 5. Degradation curves of ABS at different heating rates β (K/min).

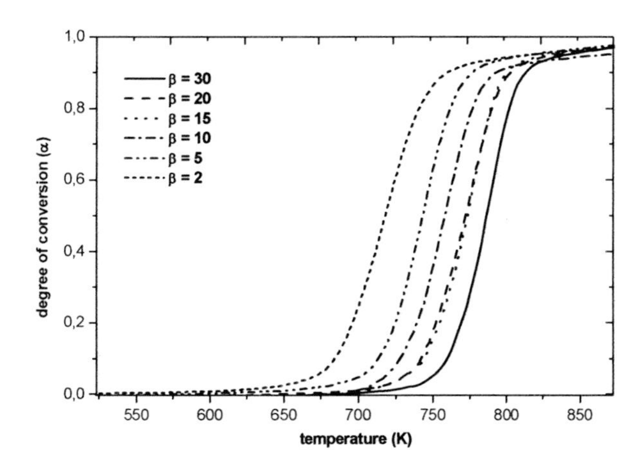

Figure 6. Degradation curves of PC at different heating rates β (K/min).

When we calculate the apparent activation energy using the classic models we obtain a very narrow variation of the parameters with the heating rate for ABS (Table 1) and a pronounced tend for PC (Table 2).

The application of the new degradation model can be carried out by using two different approaches. It is possible to calculate the model parameters (Ea, ln A) from equation (21) in a very simple way by using conventional mathematical software applications which allow us to obtain all these parameters by an iterative process. However, a simple method, based on elemental numerical analysis, can be proposed, looking for an expression that could yield a good fit to experimental results. The first mathematical operation would be a variable change:

$$\lambda(T)=K(T)t$$
$$\alpha=1-e^{-\lambda(T)_1}\left(1+\lambda(T)\right) \tag{22}$$

Table 1. Apparent activation energies, Ea (kJ/mol) for ABS at different heating rates by different classical methods.

Heating rate (K/min)	Friedman	Horowitz	Coats	Van Krevelen
2	167	176	164	234
5	161	175	159	233
10	163	186	164	251
15	157	190	152	252
20	163	190	158	249
25	163	189	156	250
30	167	187	157	247

Table 2. Apparent activation energies, Ea (kJ/mol) for PC at different heating rates by different classical methods.

Heating rate (K/min)	Friedman	Horowitz	Coats	Van Krevelen
2	171	181	184	252
5	205	228	229	324
10	223	235	235	333
15	232	280	250	378
20	250	251	256	356
25	286	272	274	392
30	308	305	293	424

This function does not take into account the K values and is independent of the curve (Figure 7) we analyze.

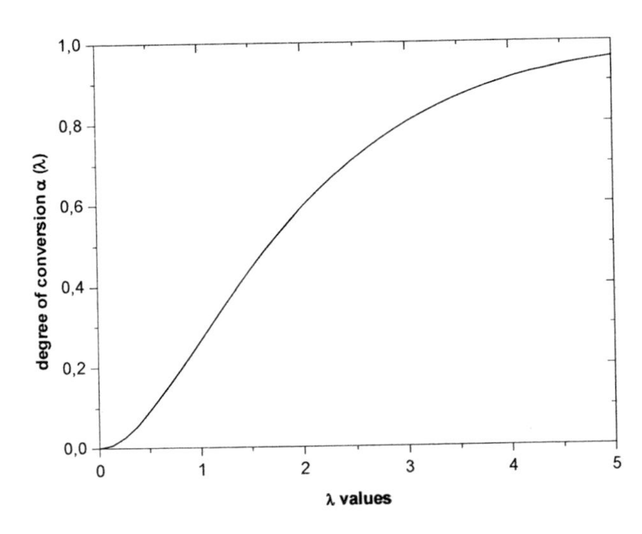

Figure 7. A sigmoidal plot which fits accurately degradation processes by a consecutive reaction model.

From the application of equation (22) we can obtain a table with some arbitrary values of λ and their corresponding α. Since the value of α can be calculated for each time and temperature from thermogravimetric data, it is possible to obtain a table with α, λ, T and t values. This will allow the calculation of the evolution of K with T throughout the whole range of α from the expression:

$$K(T) = \frac{\lambda(T)}{t} \tag{23}$$

The logarithmic representation of K(T) will allow us to obtain Ea and ln A. From these values we can simulate the curve with respect to time and temperature, although this second possibility is most commonly used. The variation of the of the K values obtained for ABS (Figure 8a) and PC (Figure 8b) with temperature shows that in both cases they follow a very similar behaviour with temperature for the different heating rates, although they show some variations in the curve shape.

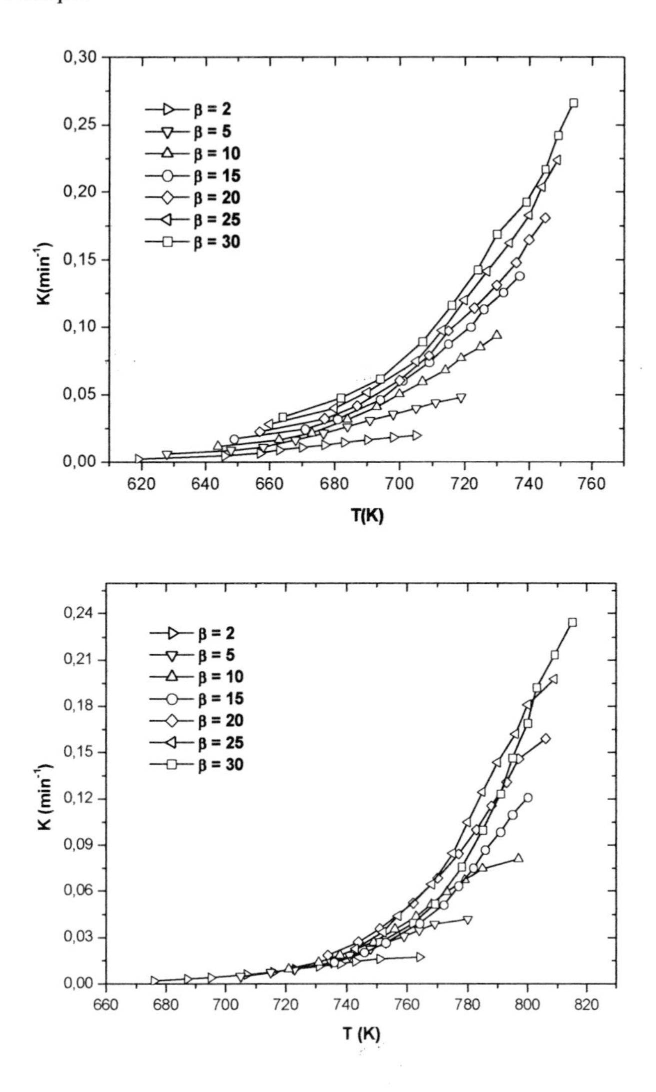

Figure 8. Variation of K values with temperature for (a) ABS and (b) PC at different heating rates β (K/min).

If we assume an Arrhenius type equation for this variation of K with T, we are able to find a good fit to ln K from α since 0.05 up to 0.9.

The most significant results obtained for both materials are indicated in Table 3. Kinetic parameters, i.e. the value of the ln A that is the intercept with the origin, and the value of the apparent activation energy that is the slope B multiplied by the constant R. The kinetic values of Ea and ln A obtained in this way follow the same trend that those obtained with other models, but they are more consistent, and show lower variation. In addition, these values allow us to reproduce quite accurately the curve from the theoretical equation with enough reproducibility and, which is more important, for the whole degradation range of both materials (Figure 9)

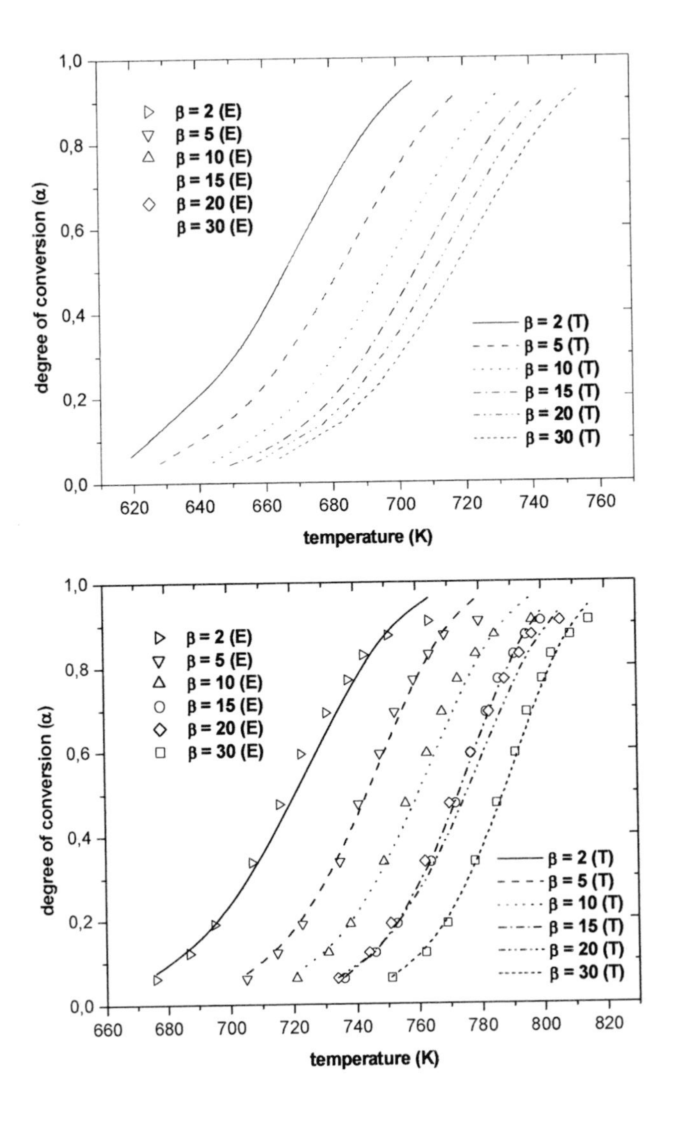

Figure 9. Comparison of theoretical (T) and experimental (E) results for (a) ABS and (b) PC at different heating rates β (K/min).

Table 3. Kinetic parameters obtained from degradation process.

Heating rate	ABS				Polycarbonate			
(K/min)	LnA	B	Ea (kJ/mol)	r	LnA	B	Ea (kJ/mol)	r
2	12,0	-11120	90,4	0,991	13,1	-12939	105,2	0,991
5	12,5	-11131	90,5	0,993	18,4	-16632	135,2	0,993
10	14,2	-12083	98,2	0,996	19,4	-17266	140,4	0,996
15	14,6	-12181	99,0	0,993	23,5	-20424	166,0	0,993
20	14,7	-12230	99,4	0,996	21,1	-18377	149,4	0,996
25	14,8	-12179	99,0	0,996	24,0	-20590	167,4	0,996
30	14,7	-12048	98,0	0,997	25,7	-21976	178,7	0,997

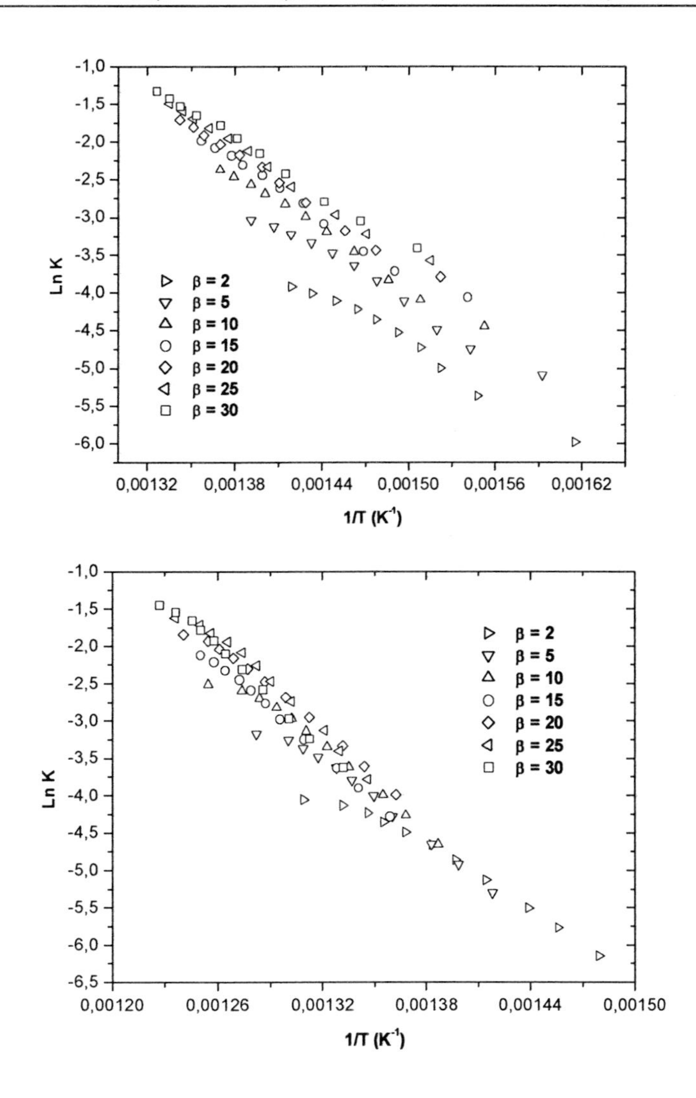

Figure 10. Arrhenius plots for (a) ABS and (b) PC at different heating rates β (K/min).

Finally, it could be interesting a graphical observation of the results of all the Arrhenius representations for all analyzed heating rates. We do not observe important differences with heating rates, even for the most extreme values. Paradoxically uniformity is higher for PC, which shows different kinetic values, than for ABS, with similar values of the kinetic parameters (Figure 10). This fact could be due to the no dependence of the degradation process with heating rate.

5. CONCLUSIONS

We have found that the classical kinetic models are still used for many degradation studies. A new model based on consecutive reactions is introduced in the present paper. When we follow the decomposition products evolution (g) in a polymer (P), we can suppose that at first moment this evolved product is close to the polymer. This compound is decomposed with a rate characterized by a K_1 constant but it disappears or volatilizes with a rate characterized by another constant K_2. Then we can follow this step experimentally in the thermogravimetric balance. The resolution of the proposed equations leads to expressions that fit reasonably well to isothermal decomposition methods.

The validation of the dynamic method for two materials, such as ABS and PC, can be considered positive. When we plot a theoretical curve from the obtained parameters, the fit is reasonably good for the whole decomposition range, not depending on the polymer. The calculated results follow the same tendency that the obtained by the application of other models, but the values are more consistent, with narrower variation.

In this model it is not necessary the incorporation of the reaction order "n". This factor is important when we work on kinetic reactions between molecules, with a physical meaning as the number of species that take part in a reaction. However, the use of reaction orders does not seem to have much sense in the study of solid decompositions, where it is just a parameter that allows us to fit different models, but without a true meaning.

ACKNOWLEDGEMENTS

Authors would like to thank Minister of Education and Science (Spain) for financial support (Project MAT2003-05511)

REFERENCES

[1] Y.N. Gupta, A. Chakraborty, G.D. Pandey, D.K. Setua, *J. App. Polym. Sci.* 92 (2004) 1737-1748.
[2] H.L. Friedman, *J. Polym. Sci. C.* 6 (1964) 183-195.
[3] J.H. Flynn, and L.A. Wall, *J. Polym. Sci. B.* 4 (1966) 323-328.
[4] W.W. Sukowski, J. Borek, A. Danch and Coll, *J. Therm. Anal. Calorim.* 77 (2004) 363-372.
[5] P. Straka, J. Nahunkova, Z. Brozova, *J. Anal. Appl. Pyrolysis*, 71 (2004) 213-221.
[6] C. Gamlin, N. Dutta, N. Roy, D. Kehoe, J. Matisons, *Thermochim. Acta* 367 (2001) 173-175.

[7] P. Budrugeac, E. Segal, L.A. Perez-Maqueda, J.M. Criado, *Polym. Degrad. Stabil.* 84 (2004) 311-320.

[8] R. Navarro, L. Torre, J.M. Kenny, A. Jiménez, *Polym. Degrad. Stabil.* 82 (2003) 279-290.

[9] D. Puglia, L.B. Manfredi, A. Vazquez, J.M. Kenny, *Polym. Degrad. Stabil.* 73 (2001) 521-527.

[10] J. M. Criado, L.A. Perez-Maqueda, in: O. Toft Sorensen, J. Rouquerol (Eds.), *Sample Controlled Termal Analysis*, Chap. 4 (2004).

[11] B.J. Holland, J.N. Hay, *Thermochim. Acta* 388 (2002) 173-175.

[12] J.J. Pysiak, Y.A. Al-Badwi, *J. Therm. Anal. Calorim.* 76 (2004) 521-528.

[13] J.A. Conesa, J.A. Caballero, J.A. Reyes, *J. Anal. Appl. Pyrol.* 71 (2004) 343-352.

[14] T. Farvelli, M. Pinciroli, F. Pisano, G. Bozzano, M. Dente, E. Ranzi, *J. Anal. Appl. Pyrol.* 60 (2001) 103-121.

[15] C. Ulloa, A. Gordon, X. Garcia, *J. Anal. Appl. Pyrol.*, 71 (2004) 465-483.

[16] J.P McCallum, J. Tanner, *Nature* 225 (1970) 1127.

[17] A.L. Drapper, *Termochim. Acta* 1(1970) 3.

[18] P.D. Garn, *J. Thermal Anal.* 6 (1976) 237.

[19] V.M. Gorbatchev, V.A. Logvinenko, *J. Thermal Anal.* 6 (1972) 237.

In: Reactions and Properties of Monomers and Polymers ISBN 1-60021-415-0
Editors: A. D'Amore and G. Zaikov, pp. 125-150

Chapter 7

SYNERGISM AND ANTAGONISM IN PROCESSES OF POLYMERS PHOTO-TRANSFORMATION AND LIGHT STABILIZATION

V. B. Ivanov
N.N. Semenov's Institute of Chemical Physics Russian Academy of Sciences
4, Kosygin str., Moscow 117977, Russia

ABSTRACT

The influence of additives nature and concentration and also of polymer structure on rates of reactions proceeding under effect of light is considered. The main attention is paid to non-additive effects observed under the use of mixtures of light stabilizers or photo-initiators in solid polymers. Theoretical dependences characteristic for the most considerable mechanisms of synergism and antagonism are analyzed. The mechanisms of influence of additives of other classes on efficiency of stabilizers and initiators action are discussed.

Keywords: light stabilizers, photo-initiators, synergism, antagonism, polymers, photo-oxidation, photo-destruction.

INTRODUCTION

Nikolai M. Emanuel paid significant attention to analysis of non-trivial effects detectable under oxidation of organic compounds and polymers. Classical N.M. Emanuel's works on conjugated oxidation of hydrocarbons [1], and also fundamental results of analysis of anti-oxidants synergism [2-6] known for author of presented article from student time to a great extent determine the theme of investigations discussed below. That interest which N.M. Emanuel showed to this problem beginning from early stages of our works directing them on solution of fundamental problems with the use of concrete practically important polymer materials as objects of investigation also served as stimulating factor.

The aim of given work is to analyze mechanisms of mutual influence of components of light stabilizing and photo-initiating systems of polymers. And the main problem is the reveling of general regularities allowing identifying the most important mechanism acting in every concrete system. The main attention was paid to quantitative aspects practically not reported in other reviews on problems of photo-destruction and light stabilization of polymers. Particularities of stabilizers action containing not only functional groups able to act by various mechanisms are also considered from these general positions.

1. SYNERGISM OF LIGHT STABILIZERS

Analysis of large number of data presenting in patent and scientific literature [7-11] on efficiency of action of light stabilizers and their mixtures and also mixtures of stabilizers with additives of other classes in various polymers allow suggesting classification of synergism mechanism taking into account the character of components mutual influence [12-14]. The main mechanisms of synergism are presented in Table 1. When discussing the importance of one or another mechanism one should consider both its experimental and theoretical validity and the value of synergism effect itself which obeys the practical criterion (1):

$$Sy = [(\tau_{1,2} - \tau_0) - (\tau_2 - \tau_0) - (\tau_1 - \tau_0)]/[(\tau_1 - \tau_0) + (\tau_2 - \tau_0)] > 0 \quad (1)$$

where $\tau_{1,2}$, τ_1 and τ_2 – the times corresponding to change of characteristic property of samples containing mixture of stabilizers ($\tau_{1,2}$) and individually taken components 1 (τ_1) or 2 (τ_2), and also non-stabilized sample (τ_0).

Diffusion mechanism of synergism is obviously the main for systems including mixtures of UV-absorbers or pigments with antioxidants. Kinetic model underlies this mechanism according to which UV-absorber (or pigment) reduces rate of consumption of antioxidant in deep layers of polymer and antioxidant protects namely the surface layer in which oxidation is preliminary initiated and proceeds. Spending of antioxidant in surface layer during the whole period of stabilizers system functioning is compensated at the expense of diffusion introduction from deep layers to a greater or lesser extent.

From analysis of general scheme of initiated polymer oxidation in the presence of stabilizers mixture general expressions were obtained characterizing light resistance in regime of fast [15, 16] and slow diffusion (in diffusion-controlled regime) [16, 17]. For stabilizers mixture including reacting with free radicals antioxidant in the regime of fast diffusion when concentration of antioxidant is practically equal in the whole sample thickness:

$$\tau = [f_1(c_{1,0} - c_1)\, \varepsilon_2 c_2 l]/w_i \quad (2)$$

and in diffusion-controlled regime:

$$\tau = [\pi D f_1(c_{1,0} - c_1)\, \varepsilon_2 c_2 l]^2/(4w_i)^2 \quad (3)$$

where $c_{1,0}$ and c_1 – initial and critical concentrations of antioxidant, f_1 and D – stoichiometric coefficient of inhibition and diffusion coefficient of antioxidant, ε_2 and c_2 – integral coefficient of extinction and concentration of UV-absorber, w_i – initiation rate, l – thickness of sample.

For antioxidant decomposing hydroperoxides without formation of free radicals in polymer oxidation of which proceeds as process with quadratic termination of chain in regime of fast diffusion:

$$\tau = \{(c_{1,0} - c_1)/k_{pr}[PH]\}f_{1.h}\varepsilon_2 c_2 l]/(k_{ter}w_i)^{1/2} \quad (4)$$

and in diffusion-controlled regime:

$$\tau = [(2\pi D k_{ter})/w_i]\{[f_{1.h}(c_{1,0} - c_1)\varepsilon_2 c_2]/(2k_{pr}[PH])\}^2 \quad (5)$$

where $f_{1.h}$ – stoichiometric coefficient of hydroperoxides decomposition by antioxidant, k_{pr} and k_{ter} – constants of rates of oxidation chain propagation and termination.

From equations (2)–(5) we may formulate the following conclusions important for practice of stabilization and prognosis of light resistance of stabilized systems:

1. Light resistance of sample in regime of fast diffusion is in proportion to screening coefficient and consequently is the higher the larger the sample's thickness;
2. The value of synergism effect depends on light intensity since at intensity increase the transition from regime of fast diffusion (when synergism is maximum) to regime of slow diffusion (when its value is decreased) and even to local regime (when synergism is absent) is possible;
3. Light resistance to a definite extent depends on molecular mass of antioxidant since the value of the last one determines its diffusion mobility in given polymer;
4. Efficiency of stabilizers mixture depends in absorbing ability of UV-absorber and soectral structure of light origin selected for accelerated tests of light resistance;
5. The maximum effect of synergism in optically thick samples of polymers is possible only at stabilizers concentrations ratio 1 : 1.

All these conclusions were experimentally confirmed [15-23]. The main part of quantitative investigations was made on rubbers or thermoelastoplasts in which diffusion of antioxidant was realized especially fast, however existing experimental data for mixtures of phenol antioxidant and UV-absorber in polypropylene [22] confirmed general character of given phenomenon. One of the illustrative examples of quantitative accordance between theory and experimental data is the dependence of light resistance on light intensity (Figure 1). It is significant that these dependences turned to be various for antioxidants interacting with free radicals (linear termination of oxidation chain) and for antioxidants decomposing hydroperoxides (at given conditions – the quadratic termination of oxidation chain).

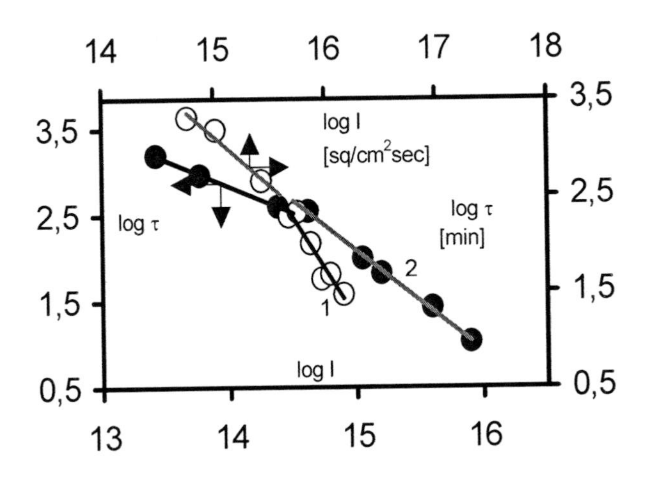

Figure 1. Synergism between UV-absorbers and antioxidants interacting with free radicals (1) and decomposing hydroperoxides without formation of free radicals (2). Dependence of induction period on light intensity under photo-oxidation of polybutadiene containing mixture of $8{,}5 \cdot 10^{-4}$ mole/kg of ether 4-hydroxy3,5-di-tret-butylphenylpropionic acid and pentaerythrite and $4{,}4 \cdot 10^{-2}$ mole/kg of 2-(2'-hydroxy-5'-methylphenyl)benzotriazole (1) [23] or isoprenstyrene thermoelastoplast containing mixture of $2 \cdot 10^{-3}$ mole/kg of nickel dibutyldithiocarbamate and $5 \cdot 10^{-3}$ mole/kg 2-(2'-hydroxy-5'-methylphenyl)benzotriazole (2) [27].

Synergism by mechanism of quenchering of excited states of antioxidant by quenchers, such as compounds of UV-absorber class, for example derivatives of 2-hydroxybenzophenone and 2-hydroxybenzotriazole also may lead to significant increase of light resistance. As it is obvious directly from its name in the base of given mechanism there is the physical phenomenon of deactivation of antioxidant molecule by quencher leading to reduction of quantum yield of antioxidant transformation and consequently also to decrease of rate of its consumption in oxidizing under the effect of light polymer.

On the base of analysis of kinetic scheme of oxidation in the presence of photo-chemically active weakly absorbing antioxidant for the value of induction period the following expression was obtained [24]:

$$\tau = \{1/[k_{ph}(1 + m_1/f_1)]\} \ln\{[w_i/f_1 + k_{ph}(1 + m_1/f_1)c_{1,0}]/[w_i/f_1 + k_{ph}(1 + m_1/f_1)c_1]\} \quad (6)$$

where k_{ph} – effective constant of rate of photo-chemical reaction of antioxidant, and m_1 – stoichiometric coefficient of formation of radicals in this reaction. At condition when initial concentration of antioxidant $c_{1,0}$ is significantly higher than critical c_1 the equation (6) is as follows:

$$\tau = \{1/[k_{ph}(1 + m_1/f_1)]\} \ln\{1 + k_{ph}(1 + m_1/f_1)c_{1,0}]/(w_i/f_1)\} \quad (7)$$

In the presence of quencher the equation (7) is transformed into equation (8):

$$\tau = \{1/[\eta k_{ph}(1 + m_1/f_1)]\} \ln\{1 + \eta k_{ph}(1 + m_1/f_1)c_{1,0}]/(w_i/f_1)\} \quad (8)$$

where n – coefficient characterizing reduction of rate of photo-transformation of anti-oxidant in the presence of quencher. At uniform distribution of additives coefficient η may be estimated by Ferster's equation [25]:

$$\eta=\varphi/\varphi_0=1 + \{(\pi/2)(c_2/c_{2,0})\exp[(\pi/4)(c_2/c_{2,0})^2]\}\{1 - \Phi[(\sqrt{\pi}/2)(c_2/c_{2,0})]\} \quad (9)$$

where $\Phi(q) = (2/\sqrt{\pi}) \int \exp(-x^2)dx$ – function of errors, c_2 – quencher concentration, $c_{2,0} = 3/(4\pi R_0^3)$, R_0 – critical distance between donor and acceptor at which energy transfer and emission are equiprobable.

The value of R_0 may be estimated from overlapping of absorption spectra of quencher and luminescence of antioxidant [26]:

$$R_0^6= \{[9000 \ln 10 K^2\varphi_0]/[128\pi^5 n^4 N_A \nu_{cp}^4]\} \int f(\nu)\varepsilon(\nu)d\nu \quad (10)$$

where ν – wave number (cm^{-1}), n – refractive exponent of medium, $\varepsilon(\nu)$ – molar coefficient of quencher absorption, $f(\nu)$ – the function of spectral distribution of donor emission normalized to 1, ν_{cp} – average wave number between maximums of donor emission and acceptor abdorption, $K^2 = 2/3$ – orientational multiplier.

They show that experimental data [27, 28] on efficiency of mixtures of spatially hindered phenols or aromatic amines with 2-(2'-hydroxy-5'-methylphenyl)benzotriazole are satisfactorily described by theoretical dependences based on calculations by equations (8)–(10).

The physical principle simple enough – the influence of additional component on stabilizers distribution – underlies one of the most significant mechanisms of synergism. It is well known [9-11, 29] that insufficient compatibility with polymer is one of the most important factors limiting efficiency of stabilizers. This disadvantage may be corrected by two methods: introduction of additional functional groups improving solubility of stabilizer in polymer [13] or usage of additives mixtures. In particular they showed that solubility of dialkyldithiocarbamate of metals was increased in both mixtures with each other [30], and with surface-active substances (SAS) [31].

The most suitable systems for analysis of synergism in stabilizers and additives systems influencing on their distribution in polymer are mixtures of light stabilizers and SAS, since SAS due to their chemical inertness should not significantly influence on light resistance of polymer [32]. Maximum effect of synergism by this mechanism as estimations of [12] show should be observed for mixtures of SAS with strongly absorbed light resistant antioxidants, since simple increase of stabilizer concentration by the value of Δc_1 is strengthened due to auto-synergism (see below) in accordance with expression (11):

$$\tau_{SAS}/\tau_0 = (c_1 + \Delta c_1)^2 /c_1^2 \quad (11)$$

where τ_{SAS} and τ_0 – the values of induction periods of photo-oxidation in the presence and absence of SAS. Detailed investigation of SAS influence on efficiency of 4-hydroxy-4'-(benzeneazo)azobenzene [33] showed that observing effects were significantly lower than predicting theoretically as a consequence of change of screening coefficient in the course of process.

Significantly larger in values synergism effects were found for mixtures of azo-compounds and spatially hindered piperidine [34]. This fact is obviously caused by both increase of solubility of stabilizers and additional effect connected with "covering" of diffusion mechanism of synergism.

The other synergism mechanisms also get definite experimental confirmations.

The most usual and trivial mechanism is formation of novel, more effective stabilizer from less effective components or from compounds not being stabilizers. One of not many examples studied in practice is formation of nickel complex from nickel nonanate and 2-hysroxy-4-methoxyacetophenonoxime in melt of polypropylene at 200°C [35]. It is important that product formed at these conditions has practically the same efficiency as preliminary synthesized complex. We naturally may consider the given system as synergetic only conditionally, since to the moment of beginning of action of main external factor–UV-light–there are no initial components in polymer, but there is only the product of their transformations.

The mechanism considering increase of sum concentration of stabilizer under the use of substances mixture each of which is restrictedly compatible with polymer is also trivial [32, 36]. Light resistance of polymer material with the use of such stabilizing systems in comparison with individual components naturally can't be increased more than in two times.

In accordance with two other mechanisms of synergism not assuming consideration of direct components interaction also, one suggests that effect of efficiency strengthening is caused by protecting of light stabilizers by antioxidant at conditions of processing [37, 38] of exploitation of polymer [39].

For mixtures of phenols with stabilizers of class of 2-hydroxybenzophenone or 2-hydroxyphenylbenzotriazole saving of stabilizers in the course of processing or exploitation is equivalent to concentration of light stabilizer by value $\mathbf{\Delta c_2}$ and increase of light resistance reaches at the expense of this fact is not higher than:

$$\Delta\tau_{1,2} \leq K\Delta c_2 = K(f_2/f_1)c_1$$

where K – proportionality coefficient the value of which depends on polymer structure, f_1 and f_2 – stoichiometry coefficients of inhibition for antioxidant and light stabilizer correspondingly, c_1 – antioxidant concentration. That is why at comparable concentrations of antioxidant and light stabilizer as it is usually in practice the increase of light resistance by considered mechanisms is determined mainly by the ratio f_1/f_2. Since the magnitudes for phenols and other hydroxyaromatic compounds are not significantly differed from 1 [13] synergism couldn't be larger in its magnitude.

Correspondingly to one more mechanism of synergism they suggest that increase of efficiency is connected with protection of polymer by additional stabilizer at stage of transition of the main component into active form [13, 40-42]. This mechanism obviously is essential under the use of mixtures of spatially hindered amines with nitroxyl radicals since amines themselves are nor stabilizers and their protecting action is connected with transformation products – nitroxyl radicals, hydroxylamines and radicals, hydroxylamines ethers [41]. Probably by this fact the synergism between phenols and derivatives of spatially hindered piperidines in polypropylene and polyethylene [43, 44] and also in other polymers

[45] is partly caused. However we couldn't exclude another mechanism considered above – the influence of phenol anti-oxidants on stages of polymers processing.

The other mechanism of light stabilizers synergism as analysis shows can't provide significant increase of light resistance.

So, fore example, the mechanism considering increase of efficiency of UV-absorber in the presence of anti-oxidant [46, 47] due to change of chain termination character leads to increase of light resistance (decrease of oxidation rate) in $\alpha_{1,2}/\alpha_2$ times calculating by the equation (12)

$$\alpha_{1,2}/\alpha_2 = \{(\lg T/\lg T_0)\,[(1 - T_0)/(1 - T)]\}/\{(\lg T/\lg T_0)\,[(1 - \sqrt{T_0})/(1 - \sqrt{T})]\} \quad (12)$$

where $\alpha_{1,2}$ – decrease of rate in the presence of UV-absorber and antioxidant (linear termination of oxidation chain); α_2 – decrease of rate in the presence of UV-absorber only (quadratic chain termination), and T_0 and T – coefficients pf samples transmission in the absence and presence of light stabilizer [13, 48]. At strong light absorption ($T \rightarrow 0$) as it is obvious from equation (12) $\alpha_{1,2}/\alpha_2 \rightarrow 2$. Actually the effect should be lower and even be not revealed at all since in the presence of UV-absorbers as weak antioxidants the photo-oxidation proceeds as process with linear chain termination [49].

Other mechanisms well known from practice of polymer thermo-stabilization including mechanism of regeneration of more effective stabilizer [1-3], influence of additional additive on critical concentration of the main stabilizer [1, 50, 51] and also interaction of additional stabilizer with catalyzing product (mainly, hydroperoxids) under photo-oxidation were not observed yet. This fact is caused mainly by photo-chemical activity of usually used anti-oxidants and also by high rates of oxidation initiation with participation of impurities and chromophorous groups [12].

Thus, synergism effects of light stabilizers may have various natures and revealed in maximum degree at various conditions. Establishing of mechanism of action of given concrete stabilizing system is possible on the base of complex analysis of synergism effect dependence on structure and also on conditions of light resistance testing and polymer properties (see Table 1). Problems appearing at that are mainly connected with the fact that in some systems several mechanisms may simultaneously act. However in this case also the contribution of each of them may be estimated by correct selection of conditions of light resistance tests and by control of physical and chemical processes proceeding with participation of stabilizers.

Table 1. The main mechanisms of stabilizers synergism.

Mechanism	Classes of stabilizing systems	Particularities of stabilizing systems	Effect at optimal conditions	Concentrations ratio	Change of effect at parameters increase			Ref #
					Sample thickness	Light intensity	Temperature	
Diffusion (antioxidants diffusion into surface layer (reaction zone))	UV-absorbers or pigments +antioxidants	Strong absorption of light stabilizer	Strong	$c_1 = c_2$	Increase	Independent or decreased	Independent or increased	[15-23]
Quenching (of excited states of antioxidants by UV-absorber)	UV-absorbers + phenols or aromatic amines	Photo-chemical activity of anti-oxidant	Moderate	$c_1 > c_2$	Independent	Independent	Independent	[27,28]
Distribution (Increase of solubility or change of stabilizer distribution)	Azo-compounds + SAS Azo-compounds + piperidines	Low solubility	Weak or moderate	$c_1 = c_2$	Independent	Independent	Independent or decreased	[32-34]
Protection under processing (protection of light stabilizer by antioxidant)	UV-absorbers + phenols	Low coefficient of inhibition	Moderate	$c_1 > c_2$	Increase	Independent or decreased	Independent or increased	[37,38]
Protection under exploitation (protection of light stabilizer by antioxidant)	UV-absorbers + phenols	Low coefficient of inhibition	Moderate	$c_1 > c_2$	Increase	Independent or decreased	Independent or increased	[39]
Protection under activation (protection of polymer by another stabilizer at stage of activation of the first one)	Piperidines+ nitroxyl radicals	Active product of stabilizer transformation	Weak	$c_1 > c_2$	Independent	Independent or decreased	Independent or decreased	[40-42]
Formation of new stabilizer (from less effective ones)	Metals salts + complexing agents	Low efficiency	Strong	$c_1 = c_2$	Independent	Independent	Independent	[35]

2. AUTO-SYNERGISM OF LIGHT STABILIZERS

In given Section auto-synergism is considered as phenomenon of self-strengthening of action of stabilizer containing various functional groups able to act by two or more various mechanisms. Existing experimental results allow marking out of two main types of this phenomenon. In accordance with the first one they assume that the main effect is caused by the fact that stabilizer simultaneously act as both UV-absorber and antioxidant and these functions are mutually independent.

With the use of conceptions on diffusion mechanism of synergism considered above for polymer photo-oxidation induction period in the presence of light-resistant strongly absorbing antioxidant the following expression was obtained [27]:

$$\tau \approx (2{,}3f\varepsilon c^2 l)/[(1 - 10^{-\varepsilon cl})w_i] \approx (2{,}3f\varepsilon c^2 l)/ w_i \quad (13)$$

As examples of light-resistant strongly absorbing antioxidants we may consider p-hydroxy- and p-aminoazo-compounds in polydienes and dienes co-polymers [34, 52] (Fig 2).

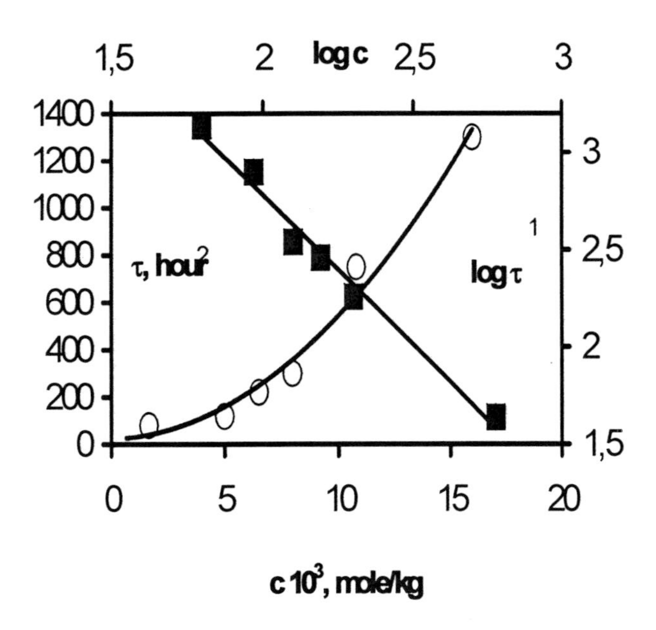

Figure 2. Auto-synergism of light-stabilizer. Increase of light resistance of isoprene-styrene thermoelastopast with the rise of concentration of 3-methyl-4hydroxy-4'-benzeneazobenzene (1) [34] and logarithmic anamorphosis of this dependence (2) in accordance to equation (13).

For these systems two the most characteristic particularities are observed:

1. Light resistance of sample (in investigated examples – the value of oxidation induction period) in accordance with equation (13) in strongly absorbing samples is in proportion to stabilizer concentration squared (Figure 2);
2. Light resistant is increased with the rise of sample thickness and at large concentrations and thicknesses dependence on thickness in accordance with formula (13) becomes linear [52].

The other mechanism assumes increase of stability to light action of group which directly functions as inhibitor at the expense of introduction of additional group providing effective deactivation of excited state [13]. Actually, in given case the other mechanism of synergism acts, i.e. that one considered above: quenchering of excited states of antioxidant by stabilizer. Obviously, for example higher efficiency of cetyl-3,5-ditretbutyl-4-hydroxybenzoate [53] in comparison with oztadecyl-3-(3',5'-ditretbutyl-4'-hydroxyphenylpropionate) is explained by this fact. In more clear form given mechanism is realized in nickel complexes containing fragments of spatially hindered phenols. In particular, nickel bis(ethyl-3,5-ditretbutyl-4-hydroxybenzyl)phosphonate is good thermo- and light-stabilizer of polyolefines [54] and polystyrene [55], whereas analogous salt of calcium is only thermo-stabilizer.

The both considered mechanisms of auto-synergism may be realized sometimes. So, for example stabilizing action of 2,4-ditretbutylphenyl-3',5'-ditretbutyl-4'-hydroxybenzoate [56] to a great extent is connected with product resulted from Fris' photo-chemical regrouping [10] – corresponding 20hydroxybenzophenone. At favourable conditions the action of this light resistant strongly absorbing antioxidant may be significantly strengthened by diffusion mechanism of synergism. High efficiency of 2-hydroxy-4-(3',5'-ditretbutyl-4'-hydroxybenzyloxy)-5-(3",5"-ditretbutyl-4"-hydroxybenzyl) benzophenone and 2-hydroxy-4-methoxy-5-(3',5'-ditretbutyl-4'- hydroxybenzyl) benzophenone combining functions of UV-absorber of 2-hydroxybenzophenone and antioxidant of class of spatially hindered phenols in thermoelastoplasts is obviously caused by simultaneous realization of both auto-synergism mechanisms [57]. Higher efficiency of molecular quenchering and consequently higher light resistance is obviously reached in calyx[n]arenas being well light stabilizers of polyvinylchloride [58].

3. Stabilizers Antagonism

Under analysis of antagonism it is advisable to issue the most general definition in accordance to which stabilizers mixtures are considered as antagonistic if their efficiency of light protecting action is lower than sum one. Effect of antagonism is as follows:

$$An = [(\tau_{1,2} - \tau_0) - (\tau_2 - \tau_0) - (\tau_1 - \tau_0)]/[(\tau_1 - \tau_0) + (\tau_2 - \tau_0)] < 0 \quad (14)$$

where values of $\tau_{1,2}$, τ_1 and τ_2 are the same as in expression (1).

In accordance with approach developed earlier [12, 13] the general property which should be taken into account for description of such systems is photochemical activity of additives. While analyzing kinetic scheme of polymer oxidation in the presence of mixture of two antioxidants differing in photo-chemical activity and inhibiting ability they showed in [13] that kinetics of antioxidants spending was determined by the following scheme of differential equations (15)–(16):

$$- dc_1/dt = f_1k_{3,1}c_1\{[w_i + m_1k_{4,1}c_1 + m_2k_{4,2}c_2]/[f_1k_{3,1}c_1 + f_2k_{3,2}c_2]\} + m_1k_{4,1}c_1 \quad (15)$$

$$- dc_2/dt = f_2k_{3,2}c_2\{[w_i + m_1k_{4,1}c_1 + m_2k_{4,2}c_2]/[f_1k_{3,1}c_1 + f_2k_{3,2}c_2]\} + m_2k_{4,2}c_2 \quad (16)$$

where $k_{3,1}$ and $k_{3,2}$ – constants of inhibition rate; $k_{4,1}$ and $k_{4,2}$ – constants of initiation rate; f_1 and f_2 – stoichiometric coefficients of inhibition; m_1 and m_2 – stoichiometric coefficients of initiation; and c_1 and c_2 – antioxidants A_1 and A_2 concentrations correspondingly.

It is clear that when one adds more photo-chemically active component, for example A_2 ($m_2k_{4,2} >> m_1k_{4,1}$) it should inevitably lead to the rise of rate of more light resistant antioxidant A_1 spending. It is clearly observed when comparing rate of spending of A_1 in the absence of A_2 (equation 17):

$$- dc_1/dt = w_i + 2m_1k_{4,1}c_1 \quad (17)$$

with sum rate of antioxidants spending (equation 18) at comparable inhibition parameters ($f_1k_{3,1} \approx f_2k_{3,2}$):

$$- (dc_1/dt + dc_1/dt) = w_i + 2m_2k_{4,2}c_2 + m_1k_{4,1}c_1 \quad (18)$$

Consequently, under addition of photo-chemically active antioxidant, as a rule decrease of light resistance should be observed. In general case at known values of constants $k_{4,1}$ and $k_{4,2}$ and also of stoichiometric coefficients f_1, f_2, m_1 and m_2 light resistance of sample in the presence of stabilizers mixture may be obtained by numerical solution of equations system (15)–(16).

Existing literature data on antagonistic systems of stabilizers may be divided into four groups in accordance with nature (principle of action) of the second component (the first component is in fact light stabilizer) (Table 2):

1. mixtures with radicals acceptors;
2. mixtures with compounds decomposing hydroperoxides without formation of free radicals;
3. mixtures with pigments and fillers;
4. mixtures with SAS.

The majority of examples of antagonism [43, 44] relates to mixtures of light stabilizers of the class of spatially hindered piperidine with spatially hindered phenols (Table 2). Obviously, the main reason of antagonism of such systems is the increase of rate of spending of piperidines photo-transformation products (nitroxyl radicals, hydroxylamines and hydroxylamine ethers) in side processes induced by phenols photo-transformation. Additional fact undoubtedly may be sensitizing action of phenols oxidation products, first of all quinines resulted from interaction of spatially hindered phenols with nitroxyl radicals in the process of processing and ageing of polymers [59-61]. In particular, the absence of correlation between efficiencies in processes of photo-and thermal oxidation testifies to importance of photo-chemical processes [43, 44]. Dependence of antagonism effect on phenols nature [62] obviously is caused by the change of their reaction ability in relation to nitroxyl radicals. Complexity of prognosis of efficiency of such stabilizers systems is caused namely by necessity of consideration of interaction between components and products of their transformation as at the stage of processing, so during exploitation of polymers.

Table 2. The main mechanisms of antagonism of polymers light stabilization.

Mechanism	Classes of stabilizing systems	Particularities of stabilizing systems	References
Photo-initiation (acceleration of spending of antioxidants due to low light stability and strong absorption of component)	Aromatic amines + phenols Phenols + metals complexes	Strong absorption. Photo-chemically activity.	[13,63-65,67]
Photo-catalysis (photo-catalytic decomposition of stabilizers by pigments)	Phenols + TiO_2. Spatially hindered amines + TiO_2 Phenols + pigments	Donor-acceptor properties of stabilizer. Photo-activity of components	[71,72,80]
Chemical interaction (of stabilizers or products of their transformation)	Phenols + nitroxyl radicals Phenols + spatially hindered amines Spatially hindered amines + sulfides Spatially hindered amines + phosphite	Chemical activity. Formation of absorbing photo-chemically active products	[43,44,59-61]
Distribution (solvation or adsorption)	Phenols + SAS Metals complexes + SAS. Azo-compounds + SAS Light stabilizers + fillers	Bad compatibility with polymer. Solvatochromism. Presence of groups able to form H-bonds with filler.	[33, 78-83]

Antagonism is also characteristic for mixtures of spatially hindered phenols with aromatic amines which noticeably inferior to individually phenols got as polybutadiene light stabilizers [63]. In these systems amines are more photo-chemically active components.

The known examples of sharp decrease of transition metals complexes efficiency in the presence of spatially hindered phenols correspond to general conceptions about antagonism of mixtures including photo-chemically active anti-oxidant [64, 65]. Under analysis of efficiency of phenols mixtures with complexes one should take into account stabilizers transformations at conditions of polymer processing [65], and also absorption of light by complexes. That is why spectral structure of light is important. Under irradiation by light with λ = 254nm the mixtures of zinc diethyldithiocarbamate with phenols possess synergetic properties [66] since complex at these conditions acts as effective UV-absorber reducing rate of phenol consumption. We should note, that correct selection of stabilizers and their concentrations allows decreasing antagonisms effects in mixtures of transition metals complexes and phenols [67].

Significant effects of antagonism were found for mixtures of spatially hindered amines and stable nitroxyl radicals with organic sulfides [41, 61, 68, 69]. The main reason of antagonism obviously is the interaction of sulfides oxidation products with amines with formation of quaternary ammonium salts and complexes with H-bond [41, 61]. Due to this the rate of amines transformation into corresponding nitroxyl radicals and other active products is sharply decreased and rate of nitroxyl radicals spending is increased [68].

The influence of spatially hindered amines of organic phosphates which as well as sulfides effectively decompose hydroperoxides without formation of free radicals on efficiency of action is less unambiguous [61, 70]. In some cases for mixtures of these stabilizers the synergism is observed [61, 70] especially noticeable under the use of aliphatic phosphates [70], however even in these systems the synergetic effect is relatively small. Antagonism strong enough is practically always characteristic for mixtures with aromatic phosphates [70] that may be connected with photo-initiating action of both phosphates themselves and the products of their transformation [70].

Pigments, as it was mentioned in Section 1 often form in mixture with antioxidants the synergetic systems acting by diffusion mechanism (see Table 20.1) since they effectively screened polymer (and antioxidant) from action of light. However in some cases, especially under the use of spatially hindered phenols as antioxidants in pigmented polymers antagonism was found [71]. In the presence of pigments antagonism between spatially hindered phenols and amines may also be strengthened [71, 72]. The presence of clear dependence of antagonism value on pigment nature and state of its surface testifies to important role of photo-catalytic processes of antioxidants transformation on the surface of pigments. Such processes are well known [73-77] and may lead to increase of initiation rate with participation of antioxidant as a result of its photo-catalytic oxidation [76, 77]. Quantitative estimations of antagonism in pigments mixtures with antioxidants are hindered due to necessity to take into account the contribution of synergism effects which are simultaneously revealed in such systems and are well observed under the use of mixtures of spatially hindered phenols with non-active pigments [71].

They also assume that phenomenon of antagonism in mixtures of light stabilizers with fillers and pigments and also the influence of fillers on antagonism effect of light stabilizing systems may be connected with adsorption of stabilizers on the surface of filler or pigment particles [61, 78-85]. Adsorption obviously is the main reason of decrease of stabilizers efficiency in the presence of photo-chemically non-active fillers such as black, chalk, talk and silica. Actually, the filler in this case serves as additive influencing on distribution of stabilizer in polymer reducing its concentration in volume where initiation is realized and polymer oxidation proceeds. In particular, the dependence of antagonism effect between black and light stabilizers of class of spatially hindered amines on black nature testifies to importance of adsorption processes [82]. Significant increase of efficiency of light stabilizers after treatment of filler (in this concrete case – talk) by reactive SAS is also the confirmation of given mechanism [83].

However in full measure the influence of filler on photo-chemical processes with participation of stabilizer obviously couldn't be totally excepted. Without consideration of this factor it is hard to explain numerous facts testifying to the presence of antagonism effects between stabilizer and filler under photo-oxidation of polymers whereas under thermo-oxidation the same systems are turned to be synergetic [79].

SAS also may function doubly in mixtures with antioxidants. In particular, they showed that fro mixtures of SAS with antioxidants of class of spatially hindered phenols in rubbers and thermoelastoplasts strong antagonism was observed [33], and the effect was the stronger the higher were the concentrations of SAS and stabilizers. Seeming contradiction between these results and those ones considered in Section 1 about synergism of mixtures of SAS with stabilizers is explained by differences in local environment of stabilizers molecules in various systems. The possible reason of efficiency decrease of polar antioxidants is their solvation in SAS micelles at concentrations higher than critical concentration of micelle formation, that leads to stabilizer molecules isolation from oxidizing under the action og light polymer. For azophenols the decrease of screening coefficient due to stabilizer solvatochromism serves as additional factor [33].

Thus, existing data on antagonism of stabilizing systems in processes of polymers photo-oxidation allow marking out of three main mechanism of this phenomenon:

1. Decrease of proper light resistance of stabilizing system under introduction of photo-chemically active additive and connected with it increase of rate of stabilizers spending. That is why this mechanism of antagonism may be conditionally called as mechanism of photo-sensitizing of light stabilizing system. Additives may act as photo-initiators directly (phenols or amines) or serve as catalyst (pigments);
2. the influence of additional component (mainly SAS and fillers) on stabilizer distribution in polymer;
3. Chemical interaction of components leading to sharp acceleration of stabilizers spending at stages of processing and exploitation. At that in some cases the products being more active photo-initiators than initial stabilizers may be formed.

Photo-chemical activity of antioxidant is the most essential factor, that is why the principles of constructing of stabilizing systems for light- and thermo-stabilization of polymers are significantly differed. As a rule, synergetic systems used under thermo-stabilization are antagonistic under photo-ageing; however antagonism effect may be significantly decreased by optimal selection of components of stabilizing system and their concentrations.

4. Auto-Antagonism of Stabilizers

Analogously to auto-synergism the auto-antagonism in given Section is considered as phenomenon of weakening of action of stabilizer containing various functional groups able to function by one and the same or various mechanisms.

This phenomenon is known long ago from the practical point of view. It was in particular shown that compounds containing two fragments of 2-hydroxy-4-alkoxybenzophenone are less active in PP than individual light stabilizers of this class [86]. Compounds containing in one molecule structures of 2-hydroxybenzophenone and spatially hindered phenol are also inactive in PP [57] that is caused by their insufficient light resistance in PP in contrast to isoprene-styrene thermoelastoplast in which stabilizers of presented type were effective enough (see Section 20.2) [57]. Low efficiency is also characteristic for compounds containing in one molecule of aromatic phosphite and spatially hindered amine or nitroxyl radical [70]. Obviously in this case also the main reason is insufficient stability of stabilizer and products of its transformation.

Low light resistance obviously is the most general reason of auto-antagonism. In full measure it relates to adducts of aromatic hydrocarbons and spatially hindered amines [87-91] which are relatively non-effective light stabilizers of PP in comparison with analogues don't containing fragments of aromatic hydrocarbons [87-90] or even not possess properties of light stabilizers at all, as for example pyrene derivatives [87].

5. Synergism of Photo-Initiators

When developing effective systems for initiation of photo-polymerization the main attention is mainly paid to increase of quantum yields of primary photo-processes [92-94] and organization of them in such way to obtain more active primary radicals able to initiate polymerization directly [92, 94]. Under creation of synergetic systems for initiation of radical processes in solid polymer matrixes naturally it is also advisable to take into account these principles. Together with them it is necessary to consider the number of specific particularities of kinetics of processes in solid solutions [7, 8], and also the considered above particularities connected with solubility and distribution of additives in polymers.

The main mechanisms of synergism of photo-initiators in slid polymers known now are presented in Table 3. Analysis of values reached with the use of synergetic systems effects testifies to the fact that the most essential mechanisms of synergism are directly connected with particularities of solid solutions.

Table 3. The main mechanism of photo-initiators synergism.

Mechansim	Classes of initiating systems	Particularities of initiating systems	Effect at optimal conditions	Concen trations ratio	Literature
Interaction with additional reagent (increase of quantum yield of primary process)	Aromatic ketones or quinines + aromatic or aliphatic amines	Low own photochemical activity of acceptor (ketone or quinine)	Moderate or weak	$c_2 > c_1$	[101-103]
Increase of radicals going out from cell (interaction with radicals of primary radical pair)	Aromatic ketones or quinines + haloid-containing compounds	Low activity of radicals of radical pair	Strong	$c_2 \approx c_1$	[99-101]
Distribution (influence of additional component on solubility or distribution of initiator in polymer)	Aromatic ketones + surface active agents	Low solubility	Moderate	$c_2 > c_1$	[106]
Interaction with additional reagent + increase of radicals goinf out from cell	Aromatic ketones or quinines + aromatic or aliphatic amines + haloid-containing compounds	Three-component systems. Low activity of ketone (quinine) and radicals of radical pair.	Strong	$c_2 \approx c_1$ $c_3 < c_1$, c_2	[101-103]
Auto-synergism (effect strengthening due to action of products of photo-initiator transformations)	Diphenylethanedion e	Decomposition with formation of acyl radicals.	Strong	-	[110]

Among the basic reasons of relatively low efficiency of photo-initiating systems action in solid polymers the most important are the following:

1. large cellular effects leading to sharp reduction of probability of radicals outlet [95, 96];
2. decrease of reaction ability of excited states in solid polymers and connected with it decrease of probability of radicals outlet [95, 96].

Decrease of cellular effects and consequently increase of radicals outlet into the volume may be reached by introduction of plasticizers and (or) temperature increase. However in practice it is difficult due to essential change of all properties of material especially its physical-mechanical characteristics. Obviously, it is more perspective to use additional reagents, fore example haloidaromatic compounds [99, 100].

Kinetic model of synergism in two-component system consisting of photo-initiator and additive increasing radicals outlet from cell at the expense of interaction with radicals of primary radical pair resulted from photo-transformation of initiator takes into account action of both two initial components and their pairs (or larger associates) [100]. In general case for rate of photo-initiation the following expression was obtained:

$$v = k_1c_1 + k_2c_2 + k_{1,2}c_{1,2} = k_1c_{1,0} + k_2c_{2,0} + (k_{1,2} - k_1 - k_2)\alpha c_{1,0}c_{2,0} + v_0 \quad (19)$$

where α – constant of distribution, v_0 – rate of process in the absence of initiators, $c_{1,0}$ and $c_{2,0}$ – concentrations of photo-initiators, introduced into polymer film, c_1, c_2 and $c_{1,2}$ – concentrations of "individual" and forming reactive pairs of molecules of photo-initiators, and k_1, k_2 and $k_{1,2}$ – effective constants of photo-chemical processes rates with participation of "free" molecules of photo-initiators and their more reactive pairs.

At constant sum concentration of photo-initiators ($c_{1,0} + c_{2,0} = c_0 = \text{const}$) expression (19) has the following form:

$$v = k_1c_{1,0} + k_2c_{2,0} + (k_{1,2} - k_1 - k_2)\alpha c_{1,0}(c_0 - c_{1,0}) + v_0 \quad (20)$$

Experimental data on radicals accumulation, oxygen absorption and formation of carbonyl groups under irradiation in vacuum or in air of films of polymethylmetacrylate or polystyrene containing mixtures of aromatic ketones or quinones (benzophenone, α-chlorantraquinone or 2,6-ditretbutylantraquinone) with α-haloidaromatic compounds (1,4-bis-trichloromethyl)benzene or 1,4-bis-(brominemethyl)benzene serve as confirmation of considered kinetic model [99-101]. In particular they showed that at low proper photo-chemical activity of components the mixtures with ratio 1 : 1 were optimal (Figure 3), and the value of synergism and its dependence on structure totally corresponded to equation (20). Under variation of second component concentration at constant content of the first one the rate of process linearly depends on concentration [99, 100].

Except considered quantitative correspondence of experimental data to theoretical dependences (19)–(20) there is a number of qualitative results confirming correctness of synergism model:

1. rates of spending of ketones or quinines in mixtures with haloidaromatic compounds are in proportion only to their proper concentration and don't depend on content of haloidaromatic compound;

2. quantum yields of processes in contrast to rates for which dependences has extreme character linearly rise with increase of concentration of haloidaromatic compound;
3. phosphorescence of aromatic ketone is weakly quenched by additives of haloidaromatic compound and dependence of relative intensity on concentration of additive has linear character;
4. under samples photolysis containing mixture of benzophenone and 1,4-bis-(trichlormethyl)benzene the evolving of HCl was registered that confirmed formation of small in sizes radicals of Cl• easily going out from cell and reacting then with matrix with formation of alkyl macroradicals.

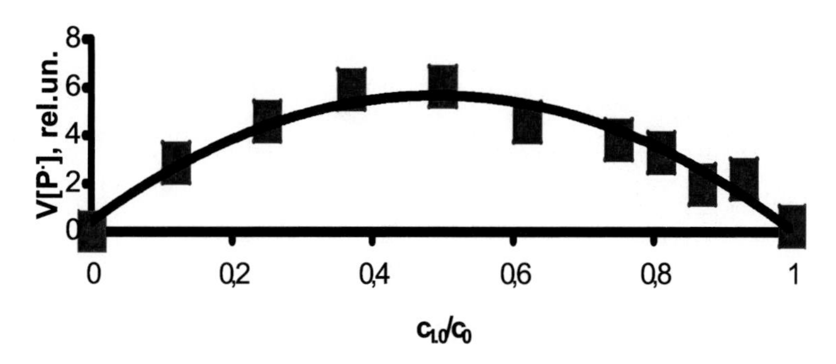

Figure 3. Synergism in two-component photo-initiating system. Dependence on initial rate of accumulation of alkyl radicals on mass part of benzophenone in mixture with 1,4-bis-(trichlorinemethyl)benzene at sum concentration 4 mass % [100]. Points characterize experimental data, and the curve presents calculation by equation (20).

Analogously to synergism at initiation of polymerization we may expect strengthening of initiation effectiveness in solid polymers and under the use of additional reagents in particular amines. Experimental data testify to the fact that it actually observed however effect of rates increase are relatively low [101-103]. Obviously, this fact is conditioned by large cellular effects since they established for model system consisting of benzophenone and diphenylamine by the method of laser photolysis that yield of radical pairs was close to 1 and their formation was accompanied by practically complete quenching of triplet states of ketones [104]. Due to low values of processes rates in two-components systems aromatic ketone (or quinine)–amine, their dependences on structure will be considered below, when analyzing synergism of three-component systems.

Under analysis of kinetic model of three-component system of photo-initiators including photo-initiator (aromatic ketone or quinine), additive increasing quantum yield of primary process (amine) and additive increasing outlet of radicals from cell (haloidaromatic compound) they used approach analogous to analysis of two-component system [101, 103]. For rate of initiation with the use of three-component system of stabilizers the following expression was obtained:

$$v = k_1c_1 + k_2c_2 + k_3c_3 + k_{1,2}c_{1,2} + k_{1,3}c_{1,3} + k_{2,3}c_{2,3} + k_{1,2,3}c_{1,2,3} + v_0 = k_1c_{1,0} + k_2c_{2,0} + K_{1,2}\alpha c_{1,0}c_{2,0} + K_{2,3}\beta c_{2,0}c_{3,0} + [K_{1,3}c_{1,0} + (K_{1,2,3}\gamma - K_{1,2}\alpha - K_{2,3}\beta)\, c_{1,0}c_{2,0}][1 - \exp(-NVc_{3,0})] + v_0 \quad (21)$$

where $K_{1,2} = (k_{1,2} - k_1 - k_2)$, $K_{1,3} = (k_{1,3} - k_1)$, $K_{2,3} = (k_{2,3} - k_2)$, $K_{1,2,3} = (k_{1,2,3} - k_1 - k_2)$, N – Avogadro constant, and index 3 relates to amine concentration or constants and associates with its participation.

When derive the equation (21) they assume that interaction of amine with aromatic ketone or quinine is described by equation (22) analogous to Perren's equation in accordance to which all molecules of ketone react with amines molecules in sphere with volume $V = (4/3)R^3$ and radius R:

$$c_{1,3} = c_{1,0}[1 - \exp(-NVc_{3,0})] \quad (22)$$

Equation (21) well describes all experimentally observed dependences of process rate on structure of photo-initiators mixtures:

1. dependence of type (22) under variation of amine (1,3,5-tromethyl-4-phenyl-4-hydroxypiperidine) concentration at constant concentrations of ketone and haloidaromatic compound in polymethylmethacrylate [102, 103];
2. linear dependence on concentration of haloidaromatic compound (1,4-bis-(trichlormethyl)benzene) at constant concentrations of ketone and amine [102, 103];
3. extreme dependence under variation of ratio between ketone and haloidaromatic compound at constant concentration of amine in polyvinylchloride [101-103];
4. extreme dependences of rates of radicals accumulation under variation of ratio of all three components in polystyrene, polymethylmetacrylate and polyvinylchloride (Figure 4), and also rate of polystyrene photo-oxidation [101-103] (Figure 5).

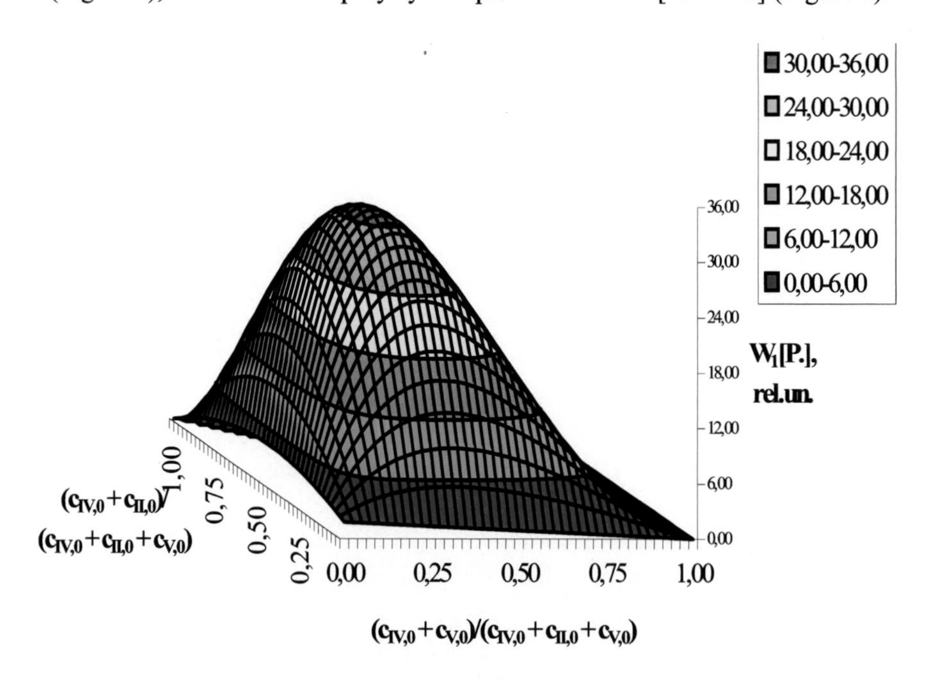

Figure 4. Synergism in three-component system. Change of rate of accumulation of alkyl radicals under variation of components ratio for mixture of 4-methoxybenzophenone (IV), 1,4-bis-(trichlormethyl)benzene (II) and N,N-dimethylaniline (III) with sum concentration 0,4 mole/kg under polyvinylchloride photolysis [101, 103]. Calculation was made by equation (21).

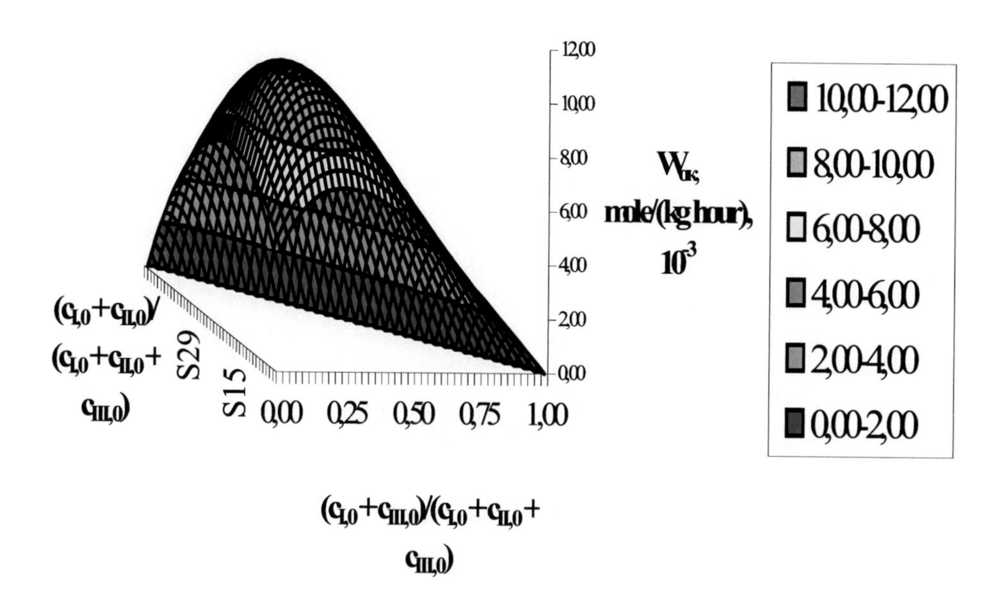

Figure 5. Synergism in three-component system. Change of polystyrene photo-oxidation rate under variation of components ratio for mixture of benzophenone (I), 1,4-bis-(trichlormethyl)benzene (II) and 1,2,5-trimethyl-4-phenyl-4-hydroxypiperidine with sum concentration 0,4 mole/kg [101, 103]. Calculation by equation (21).

As a confirmation of analyzed kinetic model of synergism (21) one considers the number of experimental facts:

1. not only qualitative, but also quantitative correspondence between theoretical dependences and experimental results. So, for example, calculated values of oxidation rates of PS containing mixture of benzophenone, 1,4-bis-(trichlormethyl)benzene and 1,3,5-trimethyl-4-phenyl-4-hydroxypiperidine are as rule differed from experimental data by not more than 10-20% [103];
2. dependence of effect of photo-initiator nature. For more active photo-initiators (for example, benzophenone) the synergism by considered mechanism is expressed in less degree than for less active (for example, for 4-methoxybenzophenone) [101, 103];
3. dependence of effect on polymer nature. For polymers with pronounced cellular effect (PS and PVC) the synergism is revealed stronger than for polymers with noticeably less in value cellular effect (PMMA) [103];
4. correspondence of determining from kinetic data values of radius of amine and ketone interaction (**R** $\approx$ 1,3nm) to the estimations obtained by quenching of phosphorescence and also proximity of this parameter to the value of molecular radius of reagents.

The majority of photo-initiators represents compounds polar enough which due to this fact are badly coincide with carbo-chain polymers. The mechanism of synergism for mixtures of photo-initiators and SAS is based on the application of known in practice of polymers light stabilization principle of increase of solubility in the presence of special additives (see Section 1) [105]. They showed that under addition of polyethyleneglycole monostearate $C_{17}H_{35}COO(CH_2CH_2O)_{35}H$ sensibilized by monoacetate α-methylbenzoyl $C_6H_5COC(OCOCH_3)(CH_2OH)C_6H_5$ the rate of oxidation of triple ethylene, propylene and dicyclopentadiene co-polymer is increased almost in order.

Detailed investigation of temperature dependences of rate of copolymer photo-oxidation and solubility of photo-initiator [105] testifies to cymbate decrease in the field of relatively low temperatures (20-90°C) under addition of SAS and also of activation energy of oxidation process (from 60 down to 30 kJ/mole), heat of photo-initiator dilution (from 120 down to 60 kJ/mole). Calculated by data on solubility of photo-uinitator values of activation energies of photo-oxidation (70 and 40 kJ/mole) are turned to be close to values obtained experimentally.

Coincidence of influence scale of SAS on solubility and rate of initiation of oxidation at one and the same temperature serves as confirmation of the supposition that action of SAS is connected mainly with the increase of photo-initiator solubility and not with the influence on value of cellular effect [105].

The other mechanism which also has its analogue among mechanisms of synergism of light stabilizers is connected with formation of new compound which is more effective photo-initiator. In particular in patent literature [106], as effective photo-initiating systems the complex compounds were suggested consisting of mono- or bicacylphosphine derivative with general formula $R_1R_2P(=O)C_6H_4R_3$ (where R_1 and R_2 – alkyl, substituted phenyl or group COR_3, R_3 – H, alkyl or alkoxyl) and α-hydroxyketone with formula $C_6H_5C(=O)C\ R_4(OH)\ R_5$ (where R_4 and R_5 – alkyl or cyclohexyl). Application of such compounds individually or in mixtures with initial components allows realization of suggested earlier approaches based on usage of mixtures of acylphosphinoxides and α-hydroxyketones obviously in optimal way [107]. In solid polymers efficiency of such complex systems is not studied however we may assume that especially for polymer they should be especially effective since they provide direct contact between components.

Analogously with considered above (Section 1) mechanism of reception of more effective light stabilizer [35] we may expect also the rise of efficiency of photo-initiation action due to formation of new photo-initiator in the course of process of polymer processing. Such effects in particular may be expected under interaction of chloric iron (representing effective but weakly absorbing in long wave region and poorly soluble photo-initiator) with sodium dialkyldithiocarbamate with formation of iron dialkyldithiocarbamate able to effective acceleration of photo-oxidation of polyolefines [108].

Essential cellular effects accompanying photo-chemical processes in solid polymer matrix lead to formation of new products not formed in liquid homogeneous solutions [109]. They show that some of these products are able to lead to strengthening of primary effects connected directly with the action of photo-initiators [110]. Such systems may be considered as synergetic and in this case system components are photo-initiator and product of its transformation

The strongest strengthening of primary effect after preliminary irradiation in adsorption band of photo-initiator was found under studying of oxidation of block-copolymer of butadiene and styrene in the presence of diphenylethanedione [110]. On the base of analysis of obtained results and literature data on thermal stability of various products [111] they assume that found effect of anomalously strong photo-chemical aftereffect is caused by free-radical canal of oxidation reaction of ketones by per-acids (Bayer-Vylliger reactions). Per-benzoic acid PhCOOOH is formed under photolysis of diphenylethandione according to the following mechanism:

$$PhCOCOPh + h\nu \rightarrow 2PhC^{\cdot}(=O)$$
$$PhC^{\cdot}(=O) + O_2 \rightarrow PhC(=O){-}OO^{\cdot}$$
$$PhC(=O){-}OO^{\cdot} + RH \rightarrow PhC(=O)OOOH + R^{\cdot}$$

Formation of radicals initiating oxidation at aftereffect stage proceeds by the following mechanism:

$$RC(=O)R + PhCOOOH \rightarrow PhC(=O){-}OO{-}CR_1R_2(OH)$$
$$PhC(=O){-}OO{-}CR_1R_2(OH) \rightarrow R_1C\,R_2(OH)O^{\cdot} + PhC(=O)O^{\cdot}$$

The most important argument in favor of suggested mechanism of photo-initiator action strengthening is coincidence of activation energy of aftereffect (E = 39,2 kJ/mole) and activation energy of peroxyether decomposition (39 kJ/mole, [111]). Besides, the mechanism confirms also the number of indirect factors [110] among which are the following:

1. Linear dependence of initial rate of photo-chemical aftereffect on dose of absorbed light;
2. Absence of effect after preliminary irradiation of samples in vacuum;
3. Decrease of effect value under addition of aromatic sulfide;
4. Presence of noticeable effect under the use of benzoyn and acetate benzoyn as photo-initiators, but complete absence of long effects of aftereffect for samples containing benzophenone.

The data presented in given Section testify to the presence of essential effects of synergism in processes of initiated oxidation of polymers. The most significant of them are connected with particularities of transformation processes of photo-initiators in solid phase among which the most important is large cellular effect caused by low molecular mobility of medium. For limitedly compatible with polymer additives the large effects may be observed also under the use of SAS increasing solubility of photo-initiators. But the largest effects obviously are reached at simultaneous action of several mechanisms mutually strengthening each other.

In given review when analyzing concrete problems of synergism under regulation of light resistance and light sensitivity of polymers the general approach was used suggested by N.M. Emanuel [8]: "Our aim is to formulate the main principles and mechanisms of phenomenon, conditions of its appearance and on this base to provide possibility of synergism prognosis and search of ways for purposeful, non-empirical or even semi-empiric selection of synergetic compositions." The same approach was used for antagonistic systems. Establishment of antagonism mechanism under polymers light stabilization is also very important since a lot of synergetic systems used in practice for increase of thermo-stabilization of polymer materials are antagonistic at light ageing. Any possibility to decrease negative consequences of this phenomenon and even more to transform antagonistic systems into synergetic represents both scientific and practical interest.

In comparison with thermal process those ones proceeding under photo-destruction and photo-oxidation of polymers have a number of principle particularities among which we want to mark two most important:

1. heterogeneity caused by prevailing localization in irradiated surface layer;
2. photo-chemical activity of the majority of used additives of thermo-stabilizers and (or) products of their transformation and also pigments.

In accordance with particularities of photo-ageing the effective stabilizer are UV-absorbers as a rule 2-hydroxyderivatives of benzophenone and phenylbenzotriazole especially in relatively thick samples. These stabilizers are often the components of synergetic systems including together with UV-absorbers antioxidants of various classes. The most effective are the mixtures of UV-absorbers with low-molecular spatially hindered amines being weak absorbing antixodants.

The main reason of essential effects of antagonism is photo-chemical activity of one or several mixture components and also the influence of additives (SAS or fillers) on stabilizers distribution. Since for complex protection of polymer material it is necessary to use effective thermo-stabilizing synergetic mixtures energies in optimization of structure of stabilizing systems should be directed on decrease of antagonism effects between light stabilizers and thermo stabilizers.

Particularities of mechanisms of synergism of photo-initiating systems are also connected with particularities of kinetics of photo-reactions in solid polymers among which the main are the low activity due to strong cellular effects and irregularity of distribution of additives into polymer volume. That is why developing of the most effective synergetic photo-initiating systems as well as synergetic light stabilizing systems should be based on fundamental principles of organic photo chemistry and kinetics of processes in solid mediums. In accordance with these principles and existing results the most perspective are the systems action of which is based on simultaneous realization of several mechanisms of synergism.

REFERENCES

[1] N.M. Emanuel, E.T. Denisov, Z.K. Maizus, *Chain reactions of hydrocarbons oxidation in liquid phase*, Moscow: Nauka (1965) *(in Russian).*

[2] N.M. Emanuel, G.V. Karpukhina, Z.K. Maizus, *Doklady AN SSSR*, 152, No.1, 110 (1963) *(in Russian).*

[3] N.M. Emanuel, G.V. Karpukhina, Z.K. Maizus, *Doklady AN SSSR*, 160, No.1, 158 (1965) *(in Russian).*

[4] N.M. Emanuel, D. Gall, *Oxidation of ehtylbenzene (model reaction)*, Moscow: Nauka (1984) *(in Russian).*

[5] G.V. Karpukhina, N.M. Emanuel, *Doklady AN SSSR*, 176, No.5, 1163 (1984) *(in Russian).*

[6] N.M. Emanuel, *Uspekhi khimii*, 54, No.9, 1393 (1985) *(in Russian).*

[7] N.M. Emanuel, A.L. Buchachenko, *Chemical physics of molecular decomposition and ageing of polymers*, Moscow: Nauka (1988) *(in Russian).*

[8] N.M. Emanuel, A.L. Buchachenko, *Chemical Physics of Polymer Degradation and Stabilization*, VNU Science Press. Utrecht (1987).
[9] *Plastics Additives Handbook*, Ed. by Zweifel H. Munich: Hanser Publ. (2001).
[10] V.Ya. Shlyapintokh, *Photochemical Conversion and Stabilization of Polymers*, Munich: Hanser Publ. (1984).
[11] N. Grassie, G. Scott, *Polymer Degradation and Stabilization*, New York e.a.: Cambridge Univ. Press (1985).
[12] V.B. Ivanov, V.Ya. Shlyapintokh, *Dev. Polym. Stab.*, 8, 29 (1987).
[13] V.B. Ivanov, L.V. Samsonova, *Itogi nauki i tekhniki. Khimiya i tekhnologiya vysokomol. soed.*, VINITI, 24, 3 (1988) *(in Russian)*.
[14] A.A. Ferasimenko, A.K. Batalov, et. al., *Protection from corrosion, ageing and bioinjuries*, Manual, Moscow: Mashinostroenie (1987) *(in Russian)*.
[15] V.B. Ivanov, N.A. Rozenboim, L.G. Angert, V.Ya. Shlyapintokh, *Doklady AN SSSR*, 241, No.3, 612 (1978) *(in Russian)*.
[16] V.B. I vanov, S.F. Burlatsky, N.A. Rozenboym, V.Ya. Shlyapintokh, *Eur. Polym. J.*, 16, No.1, 65 (1980).
[17] V.B. Ivanov, E.L. Lozovskaya, *Works of Moscow physical-technical institute. General and molecular physics*, 77 (1980) *(in Russian)*.
[18] N.A. Rozenboim, L.G. Angert, V.B. Ivanov, V.Ya. Shlyapintokh, *Kauchuk i rezina*, No.2, 31 (1980) *(in Russian)*.
[19] V.B. Ivanov, E.L. Losovskaya, A.F. Efremkin, V.Ya. Shalyapintokh, *Angew. Makromol. Chem.*, 114, No.1, 35 (1983).
[20] E.L. Losovskaya, V.Ya. Shlyapintokh, *Polym. Photochem.*, 3, No.3, 235 (1983).
[21] V.Ya. Shlyapintokh, *Pure and Appl. Chem.*, 55, N.10, 1661 (1983).
[22] A.A. Efimov, V.B. Ivanov, G.V. Kutimova, E.L. Losovskaya, V.Ya. Shlyapintokh, *Polym. Photochem.*, 3, No.4, 231 (1983).
[23] V.B. Ivanov, N.A. Rozenboim, L.G. Angert, *Vysokomol. Soed.*, 24B, No.3, 324 (1982) *(in Russian)*.
[24] V.B. Ivanov, A.F. Efremkin, N.A. Rozenboim, V.Ya. Shlyapintokh, *Vysokomol. Soed.*, 25A, No.6, 1209 (1983) *(in Russian)*.
[25] Th. Forster, *Zeit. Naturforschung.*, 4A, No.2, 321 (1949).
[26] Th. Forster, *Disc. Faraday Soc.*, 27, No.1, 7 (1959).
[27] E.L. Lozovskaya, V.B. Ivanov, V.Ya. Shlyapintokh, *Vysokomol. Soed.*, 27A, No.8, 1589 (1985) *(in Russian)*.
[28] E.L. Lozovskaya, *Diss. of candidate of chemical sciences,* Moscow: IKhF RAS (1984) *(in Russian)*.
[29] B. Renbi, Ya. Rabek, *Photo-destruction, photo-oxidation and photo-stabilization of polymers*, Moscow: Mir (1978).
[30] A.A. Dontzov, *Processes of elastomers structurization*, Moscow: Khimiya (1978) *(in Russian)*.
[31] Z.N. Tarasova, A.A. Dontzov, V.A. Shershnev, I.D. Khodzhaeva, *Kauchuk i rezina*, No.6, 16 (1977) *(in Russian)*.
[32] A.F. Efremkin, V.B. Ivanov, *Vysokomol. Soed.*, 24B, No.8, 622 (1982) *(in Russian)*.
[33] A.F. Efremkin, V.B. Ivanov, *Kauchuk i rezina*, No.9, 17 (1987) *(in Russian)*.
[34] A.F. Efremkin, V.B. Ivanov, *Polym. Photochem.*, 4, No.3, 179 (1984).
[35] N.S. Allen, J. Homer, J.F. McKellar, *J. Appl. Polym. Sci.*, 22, No.3, 611 (1978).

[36] V.B. Ivanov, *Diss. of Doctor of chemical sciences*, Msocow: IKhF (1985) *(in Russian)*.
[37] G. Scott, *Dev. Polym. Stab.*,. 6, 29 (1983).
[38] G. Scott, M.F. Yusoff, *Eur. Polym. J.*, 16, 497 (1980).
[39] G. Scott, M.F. Yusoff, *Polym. Degrad. Stab.*, 2, No.3, 309 (1980).
[40] V.B. Ivanov, V.Ya. Shlyapintokh, *Polym. Degrad. Stab.*, 28, No.3, 249 (1990).
[41] V.Ya. Shlyapintokh, V.B. Ivanov, *Dev. Polym. Stab.*, 5, 41 (1982).
[42] V.Ya. Shlyapintokh, V.B. Ivanov, O.M. Khvostach, A.B. Shapiro, E.G. Rozantzev, *Doklady AN SSSR*, 225, No.5, 1132 (1975) *(in Russian)*.
[43] N.S. Allen, A. Hamidi, F.F. Loffelman, P. MacDonald, M. Rauhut, P.V. Susi, *Plast. Rubb. Process. Appl.*, 5, No.3, 259 (1985).
[44] N.S. Allen, A. Hamidi, D.A. Williams, F.F. Loffelman, P. MacDonald, P.V. Susi, *Plast. Rubb. Process. Appl.*, 6, No.2, 109 (1986).
[45] P. Caucik, M. Povazankova, J. Durmis, M. Karvas, *Angew. makromol. Chem.*, 137, 249 (1985).
[46] O.N. Karpukhin, *Mater. Plast. Elast.*, 3, No.2, 116 (1976).
[47] Yu.A. Ershov, G.P. Gladyshev, *Vysokomol. Soed.*, 27A, 1267 (1977) *(in Russian)*.
[48] A.L. Margolin, V.A. Velichko, A.V. Sorokina, L.M. Postnikov, V.S. Levin, M.Ya. Zabara, V.Ya. Shlyapintokh, *Vysokomol. Soed.*, 27A, No.6, 1313 (1985) *(in Russian)*.
[49] P. Vink, *Degradation and Stabilization of Polyolefins*, Ed. by Allen N.S. London: Appl. Sci. Publ. (1983).
[50] Yu.A. Shlyapnikov, *Dev. Polym. Stab.*, 5, 1 (1982).
[51] Yu.A. Shlyapnikov, S.G. Kiryushkin, A.P. Mar'in, *Anti-oxidative polymer stabilization*, Moscow: Khimiya (1986) *(in Russian)*.
[52] V.B. Ivanov, A.F. Efremkin, L.V. Arinich, M.V. Gorelik, V.Ya. Shlyapintokh, *Izv. AN SSSR. Ser. Khim.*, No.9, 2019 (1982) *(in Russian)*.
[53] J.A. Stretansky, *Soc. Plast. Eng. Reg. Tech. Conf. Houston.*, Houston (1981).
[54] N.S. Allen, A. Chirinos-Padron, J.H. Appleyard, *Polym. Degr. Stab.*, 5, No.5, 323 (1983).
[55] M. Novakovska, *Polym. Photochem.*, 3, No.4, 243 (1983).
[56] J.A. Mock, *Plast. Eng.*, 38, No.4, 35 (1982).
[57] N.A. Mukmeneva, S.V. Bukharov, G,N, Nugumanova, N.V. Lysun, V.B. Ivanov, *Vysokomol. Soed.*, 40B, No.9, 1506 (1998) *(in Russian)*.
[58] A. Ninagawa, K. Cho, H. Matsuda, *Makromol Chem.*, 186, No.7, 1379 (1985).
[59] N.S. Allen, *Macromol. Chem. Rapid Commun.*, 1, No.4, 245 (1980).
[60] N.S. Allen, *New Trends Photochem. Polym. Proc. Int. Symp. hon. Prof. Bengt Ranby.*, London, New York. (1985).
[61] K. Kikkawa, *Polym. Degrad. Stab.*, 49, No.1, 135 (1995).
[62] J. Lucki, J. Rabek, B. Ranby, *Polym. Photochem.*, 5, No.1-6, 351 (1984).
[63] M.-A. De Paoli, G.W. Shulz, L.T. Furbau, *J. Appl. Polym. Sci.*, 29, No.8, 2493 (1984).
[64] S. Al-Malaika, G. Scott, *Eur. Polym. J.*, 19, No.3, 241 (1983).
[65] S. Al-Malaika, K.B.Chakraborty, G. Scott, *Dev. Polym. Stab.*, 6, 73 (1983).
[66] E.G. Kolawole, *J. Appl. Polym. Sci.*, 27, No.19, 3437 (1982).
[67] V.P. Kolomytzyn, A.A. Efimov, Z.G. Popova, P.F. Ivanenko, M.P. Zverev, *Khim. volokna*, No.5, 10 (1984) *(in Russian)*.
[68] J. Lucki, S.Z. Jian, J.F. Rabek, B. Ranby, *Polym. Photochem.*, 7, No.1, 27 (1986).
[69] K. Kikkawa, Y. Nakahara, Y. Ohkatsu, *Polym. Dedrad. Stab.*, 18, No.2, 237 (1986).

[70] I. Bauer, W.D. Habicher, S. Korner, S. Al-Malaika, *Polym. Degrad. Stab.*, 55, No.2, 217 (1997).
[71] N.S. Allen, J.L. Gardette, J. Lemaire, *Dyes and Pigm.*, 3, No.4, 295 (1982).
[72] N.S. Allen, A. Parkinson, *Polym. Degrad. Stab.*, 5, No.3, 189 (1983).
[73] P.P. Klemchuk, *Polym. Photochem.*, 3, No.4, 1 (1983).
[74] K. Okamoto, Y. Yamamoto, H. Tanaka, M. Tanaka, A. Itaya, *Bull. Chem. Soc. Jap.*, 58, No.7, 2015 (1985).
[75] K. Okamoto, Y. Yamamoto, H. Tanaka, A. Itaya, *Bull. Chem. Soc. Jap.*, 58, No.7, 2023 (1985).
[76] D. Vione, T. Picatonotto, M.E. Carlotti, *J. Cosmet. Sci.*, 54, No.5, 513 (2003).
[77] D. Bahnemann, *The Handbook of Environmental Chemistry. Vol. 2. Part 1. Environmental Photochemistry.*, Ed. by Boule R. Berlin. Springer-Verlag. (1999).
[78] H. Xingzhou, X. Hongmei, Z. Zhefeng, *Polym. Degrad. Stab.*, 43, No.2, 225 (1994).
[79] N.S. Allen, M. Edge, T. Corrales, A. Chiolds, C. Liauw, F. Catalina, C. Peinado, A. Minihan, *Polym. Degrad. Stab.*, 56, No.2, 125 (1997).
[80] N.S. Allen, M. Edge, T. Corrales, F. Catalina, *Polym. Degrad. Stab.*, 61, No.1, 139 (1998).
[81] C.M. Liauw, A. Childs, N.S. Allen, M. Edge, K.R. Franklin, D.G. Collopy, *Polym. Degrad. Stab.*, 65, No.2, 207 (1999).
[82] J.M. Pena, N.S. Allen, M. Edge, C.M. Liauw, B. Valange, *Polym. Degrad. Stab.*, 72, No.2, 259 (2001).
[83] P. Anna, G. Bertalan, G. Marosi, I. Ravadits, M.A. Maatoug, *Polym. Degrad. Stab.*, 73, 463 (2001).
[84] *Handbook of Polymer Degradation*, Ed. By Hamid S.H. New York: Marcel Dekker (2000).
[85] D. Feldman, *J. Polym. Environ.*, 10, No.4, 163 (2002).
[86] Yu.I. Temchin, E.F. Burmistrov, L.A. Skripko, R.S. Burmistrova, Yu.V. Kokhanov, M.A. Gushina, E.G. Rozantzev, *Vysokomol. Soed.*, 15, No.5, 1038 (1973) *(in Russian).*
[87] P. Hrdlovic, S. Chmela, *Polym. Degrad. Stab.*, 61, No.2, 177 (1998).
[88] S. Chmela, M. Danko, P. Hradlovich, *Polym. Degrad. Stab.*, 63, No.1, 159 (1999).
[89] R.A. Ortiz, E.A.R. Salas, N.S. Allen, *Plolym. Degrad. Stab.*, 60, No.1, 195 (1998).
[90] R.A. Ortoz, A.E. Lara, N.S. Allen, *Polym. Degrad. Stab.*, 64, No.1, 49 (1999).
[91] P. Hrdlovoc, M. Danko, S. Chmela, *J. Photochem. Photobiol. A: Chem.*, 149, No.2, 207 (2002).
[92] B.M. Monroe, *Chem. Rev.*, 93, No.2, 435 (1993).
[93] A. Costela, I. Garcia-Moreno, O. Garcia, R. Sastre, *J. Photochem. Photobiol. A.: Chem.*, 131, No.2, 133 (2000).
[94] M.V. Ecinas, A.M. Rufs, S. Bertolotti, C.M. Previtali, *Macromolecules*, 34, No.8, 2845 (2001).
[95] P.Yu. Butyagin, V.B. Ivanov, V.Ya. Shlyapintokh, *Plast. massy*, No.7, 15 (1979) *(in Russian).*
[96] P.P. Levin, V.A. Kuzmin, V.B. Ivanov, V.V. Selikhov, *Izv. AN SSSR. Ser. Khim.*, No.8, 1742 (1989).
[97] V.V. Selikhov, V.B. Ivanov, *Izv. AN SSSR. Ser. Khim.*, No.1, 18 (1990) *(in Russian).*
[98] A.A. Kachan, P.V. Zamotaev, *Photo-chemical modifying of polyolefines*, Kiev: Naukova Dumka (1990) *(in Russian).*

[99] V.B. Ivanov, E.Yu. Khavina, *Plast. massy*, No.1, 35 (1998) *(in Russian)*.

[100] V.B. Ivanov, E.Yu. Khavina, *Vysokomol. Soed.*, 41A, No.7, 1 (1999) *(in Russian)*.

[101] V.B. Ivanov, E.Yu. Khavina, *Aging of Polymers, Polymer Blends and Polymer Composites*, Ed. by G.E. Zaikov, A.L. Buchachenko, V.B. Ivanov, New York. Nova Sci. Publ. (2002).

[102] V.B. Ivanov, E.Yu. Khavina, *Plast. massy*, No.12, 23 (2000) *(in Russian)*.

[103] V.B. Ivanov, E.Yu. Khavina, *Vysokomol. Soed.*, 44A, No.12, 2084 (2002) *(in Russian)*.

[104] V.B. Ivanov, P.P.Levin, V.V. Selikhov, E.Yu. Khavina, *Doklady AN SSSR*, 367, No.2, 204 (1999) *(in Russian)*.

[105] A.F. Efremkin, A.P. Mar'in, V.N. Ovchinnikov, V.B. Ivanov, *Vysokomol. Soed.*, 30B, No.9, 710 (1988) *(in Russian)*.

[106] Patent 2181726 Russia (2002).

[107] Patent 2259704 UK (1993).

[108] G. Scott, *Dev. Polym. Stab.*, 8, 209 (1987).

[109] A.F. Efremkin, V.B. Ivanov, *Izv. AN SSSR. Ser. Khim.*, No.6, 1419 (1996) *(in Russian)*.

[110] V.B. Ivanov, E.Yu. Khavina, *Vysokomol. Soed.*, 40A, No.2, 225 (1998) *(in Russian)*.

[111] T.F. Shumkina, S.G. Voronina, A.L. Perkel', *Zh. Prikl. Khimii*, 69, No.2, 287 (1996) *(in Russian)*.

In: Reactions and Properties of Monomers and Polymers ISBN: 1-60021-415-0
Editors: A. D'Amore and G. Zaikov, pp. 151-163

Chapter 8

OZONOLYSIS OF POLYBUTADIENES WITH DIFFERENT MICROSTRUCTURE IN SOLUTION

M. P. Anachkov, S. K. Rakovsky and G. E. Zaikov*
Institute of Catalysis, Bulgarian Academy of Sciences,
Acad. G. Bonchev bl.11, 1113 Sofia, BULGARIA
*Institute of Biochemical Physics Russian Academy of Sciences,
4, Kosygin Street, Moscow 119991 (Russia)

ABSTRACT

The reaction of ozone with 1,4-cis-polybutadiene (E-BR) and polybutadiene having the following linking of the butadiene units (BR): 1,4-cis (47%), 1,4-trans (42%), 1,2 (11%) was investigated in CCl_4 solution. It was determined by means of IR-spectroscopy and ^{1}H-NMR spectroscopy that the basic ozonolysis products of both elastomers are ozonides and aldehydes. The aldehyde:ozonide ratio was 11:89 and 27:73 for E-BR and BR, respectively. In addition, epoxide groups were detected only in the case of BR, their yield was about 10 % of that of the aldehydes. Based on BR ozonolysis it was established that the ozonide yield from 1,4-trans units is about 50 %. By using the aldehyde yields, an evaluation was made of the efficiency of ozone degradation of the two polybutadienes, according to which the respective value of BR is considerably higher than that of E-BR. A reaction mechanism is proposed, which explains the formation of the identified functional groups and the differences in the ozone degradation of the studied elastomers.

Keywords: Ozonolysis, Polybutadiene, Solution, Ozone degradation.

INTRODUCTION

It is well known that there exist some differences in the kinetics and in the mechanism of ozonolysis of the respective cis and trans alkenes [1-3]. It is established that as a rule the values of the rate constants for the trans-alkenes are about 1.5-2 times higher than those of the cis-forms [4-9]. On the other side the yield of the basic reaction product in the so called "non-participating solvents" – ozonide is usually greater in the case of cis-alkenes [2, 10, 11]. The ozone reaction with polydienes in solution is used for determination of the elastomers structure [12, 13], their modification [14] and especially for investigations on the kinetics and mechanism of ozone degradation [15, 16]. The degradation, observed in solution, is due to the proceeding of non-ozonide routes of ozonolysis, which are accompanied by formation of terminal carbonyl groups [17-19]. Our comparative study of the ozonolysis of 1,4-cis and 1,4-trans – polyisoprenes did not find considerable differences in the yields of ozonides and carbonyl compounds, and in the efficiency of ozone degradation of the two elastomers [20].

In the present paper an attempt is made to examine some peculiarities of the ozonolysis of cis- and trans- polymeric double bonds of two polybutadienes with different microstructure.

EXPERIMENTAL

Materials

Ozone. Ozone was prepared by passing oxygen through a 4-9 kV discharge.

Rubbers. Commercial samples of 1,4-cis-polybutadiene rubber (SKD) and polybutadiene (Diene 35 NFA) were used in the experiments. The molecular masses of the elastomers were determined by the GPC method on a Waters instrument. The following values of weight and number average molecular masses, respectively, were obtained: 413 000 and 154 700 for the SKD (E-BR)-samples; 298 000 and 113 300 for the Diene 35 NFA (BR) [21]. All rubbers were purified by a threefold precipitation from CCl_4 solutions with excess of methanol.

The microstructure of the polymers (in %), determined from respective ^{1}H-NMR spectra, was as follows SKD (1,4-cis->96); Diene 35 NFA (1,4-cis – 47; 1,4-trans – 42; 1,2 – 11).

Procedures and Apparatus

Ozonation of polymer solutions. The experiments were performed in a bubbling reactor, equipped with a thermostatic water jacket and a viscometer. An ozone-oxygen mixture with ozone concentration in the $(0.1\text{-}1.2)\text{x}10^{-4}$ mol/litre range was passed through the reactor at a rate of $(1.5\pm0.15)\text{x}10^{-3}$ litre/s. The ozone concentrations in the gas phase at the reactor inlet and outlet were measured spectrophotometrically at 254 nm [3]. The volume of the polymer solutions was 15 ml for the viscometric and 10 ml for the rest of measurements. The changes in the molecular mass during ozonolysis were registered viscometrically. The algorithm for determination of the characteristic viscosity and of the degradation efficiency, defined by the number of viscometrically detected chain scission acts with respect to one molecule of reacted ozone, is described elsewhere [22]. All the viscometric measurements were performed at 25 ^{o}C.

Infrared spectra were acquired on a Carl Zeiss Specord IR-71 dispersive spectrophotometer. *1H-nuclear magnetic resonance spectra* were recorded on a Bruker WM 250 MHz instrument.

RESULTS AND DISCUSSION

Viscometric Study of Ozone Degradation

The ozone bubbling through polybutadienes solutions results in a decrease in their viscosity and, thus, in the corresponding molecular masses (Figure 1). It is seen that the degradation of BR is more intensive. The number of chain scissions per molecule of reacted ozone (φ), derived from Figure 1, was used to measure the efficiency of degradation of the two polybutadies towards the consumed ozone.The obtained values at $[O_3] = 1.2\ 10^5$ mol/l for SKD and Diene 35 NFA were 0.007 and 0.009, respectively. It has been found that at a degree of ozone conversion of the double bonds higher than 20-25 % a gel formation is registered [22]. For that reason the ozone degradation of 1,4-cis-polybutadiene has been described as a competitive process of degradation and crosslinking routes [22]. However, it has been determined that the relative part of the crosslinking reactions in the overall balance of the consumed ozone is very low compared with the share of the degradation routes, especially under conditions of relatively low values of $[O_3]$ [17].

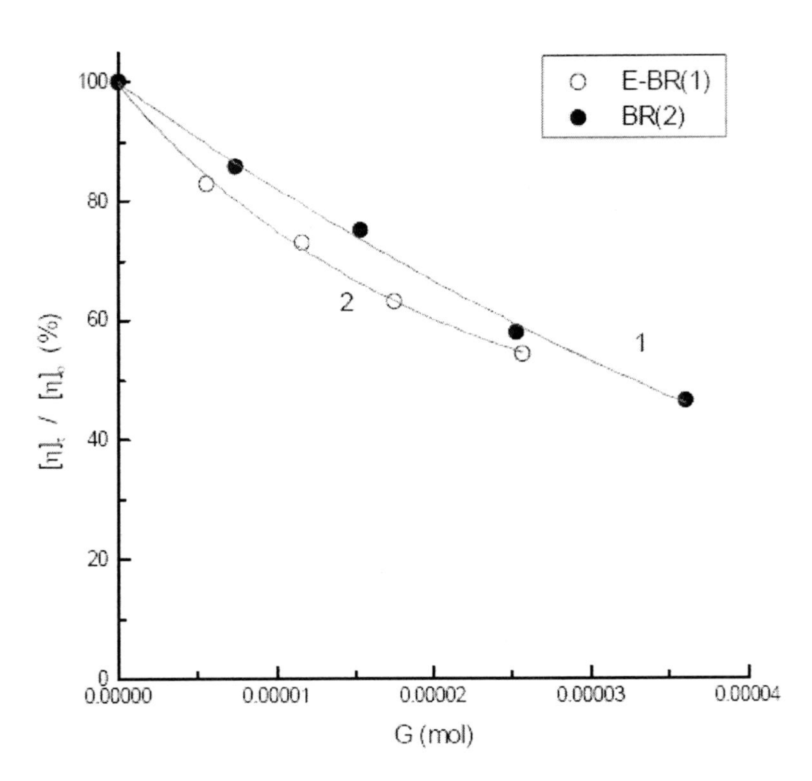

Figure 1. Dependence of $[\eta]_t / [\eta]_o$ (%) on the amount of reacted ozone (G) for carbon tetrachloride solutions (0.6 g/100 ml) of E-BR and BR. Note that $[\eta]_o$ and $[\eta]_t$ are the intrinsic viscosity values prior to and during ozonization, respectively; ozone concentration at the reactor inlet was $5x10^{-5}$ mol/litre.

The φ values, determined by viscometric measurements, do not correspond to the real extent of the degradation efficiency of the elastomers [16, 19, 22, 23]. The discrepancy between the viscometric data and those obtained by functional group analysis is a result of uneven distribution of the degradation process in the bubbling reactor. The characteristics of the mass transfer of reagents during ozonolysis in bubbling reactor were discussed in detail in Ref. [16, 22]. In the case of 1,4-cis-polybutadiene, because of the very high values of its rate constant with ozone (6.10^4 $l.mol^{-1}.s^{-1}$ [2]), the instantaneous volume of the polymeric solution layer, in which the reaction takes place is about 1.1 10^{-3} cm^3 (0.007 % from the total volume) [23]; the frequency of replacement of the boundary surface of the liquid phase is about 250 times per second [22]. The intensive progress of the reaction in a very small volume determines the formation of significant amount of low molecular mass fragments which in practice cannot be viscometrically detected.

The examination of the differences in the φ values for E-BR and BR should be based mainly on two factors: (i) the conditions of mass transfer of the reagents, and (ii) the reaction mechanism. Under other identical conditions the mass transfer parameters are determined by the rate of the reaction. It is usually assumed that the 1,4-trans double bonds react with ozone about two times faster than the respective 1,4-cis-double bonds [3-9]. In this connection the rate of the ozone reaction with BR can be evaluated based on the already mentioned suggestion about respective k value and 42 % content of the 1,4-trans double bonds. Our estimation is that the influence of the relatively small difference in the rate values of E-BR and BR could not be significant. As far as the reaction mechanism is concerned, it will be shown further that in the case of BR, the relative share of the routes responsible for degradation in the overall balance of consumed ozone, is considerably greater, in comparison with the respective one of E-BR.

Functional Group Analysis

IR-spectra of non-ozonized and ozonized polybutadiene solutions are shown in Figures 2 and 3. In the spectra of the ozonized polybutadienes the appearance of bands at 1111 и 1735 cm^{-1}, that are characteristic of ozonide and aldehyde groups, corespondingly, is observed [2, 3, 17]. It was found out that the integral intensity of ozonide peak in the E-BR spectrum, is greater and of the aldehyde one is considerably smaller in comparison with the respective peaks in the BR spectrum, at one and the same ozone conversion degree of the double bonds. The mentioned differences in the aldehyde yields indicate that, according to IR-analysis, the degradation efficiency of the BR solutions is greater.

The ^{1}H-NMR spectroscopy provides much more opportunities for identification and quantitative determination of functional groups, formed during ozonolysis of polybutadienes. Figures 4 and 5 show spectra of ozonized E-BR and BR, respectively. The signals of the ozonolysis products are decoded in Table 1 on the basis of Fig. 6. The ozonide: aldehyde ratio, determined from NMR spectra, was 89:11 and 73:27 for E-BR and BR, correspondingly. The peak at 2.81 ppm is present only in the spectra of ozonized Diene 35 NFA. It is usually associated with the occurrence of epoxide groups [26]. The integrated intensity of that signal compared to the signal of aldehyde protons at 9.70-9.79 ppm was about 10 %. Similar signal at 2.75 ppm has been registered in the spectra of ozonized butadiene-nitrile rubbers, where the 1,4-trans double bonds are dominant [23].

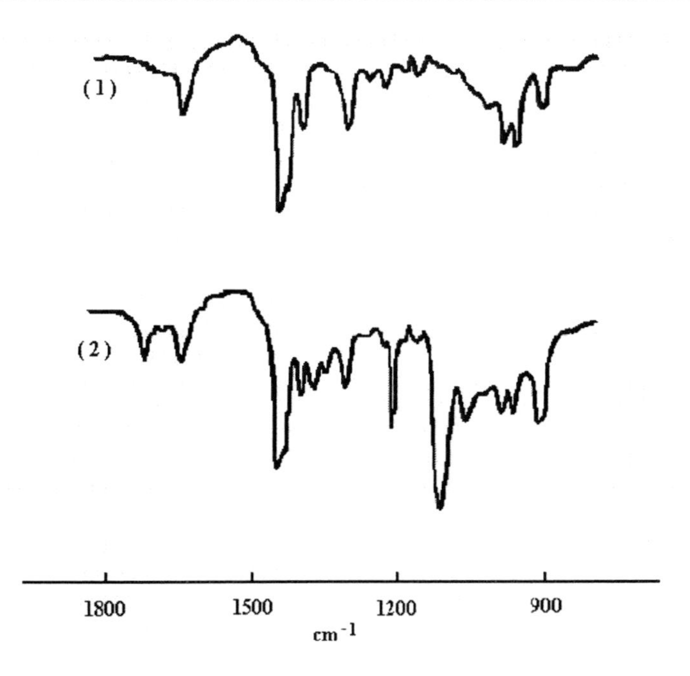

Figure 2. IR-spectra of E-BR solutions (0.89 g / 100 ml CCl_4): (1) non-ozonized; (2) ozonized to 18 % conversion of the double bonds.

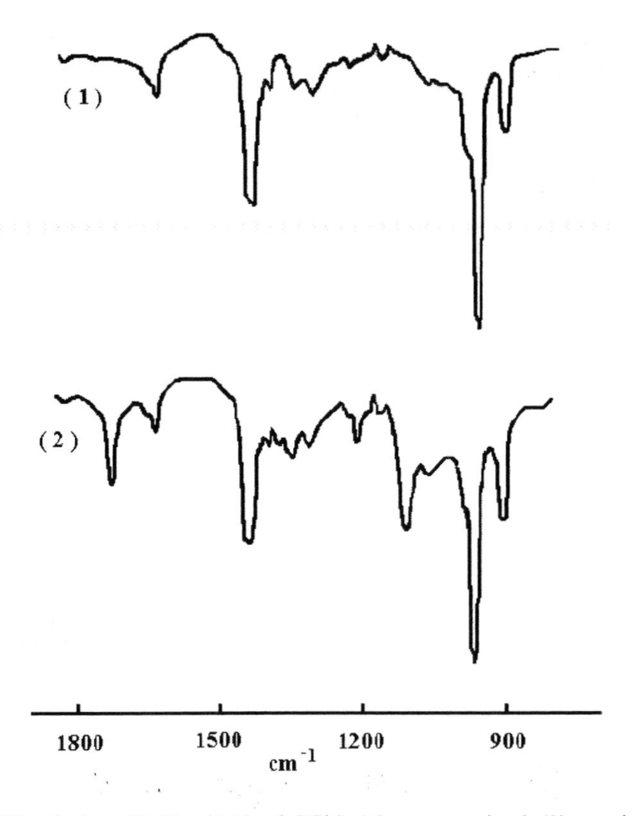

Figure 3. IR-spectra of BR solutions (0.89 g / 100 ml CCl_4): (1) non-ozonized; (2) ozonized to 18 % conversion of the double bonds.

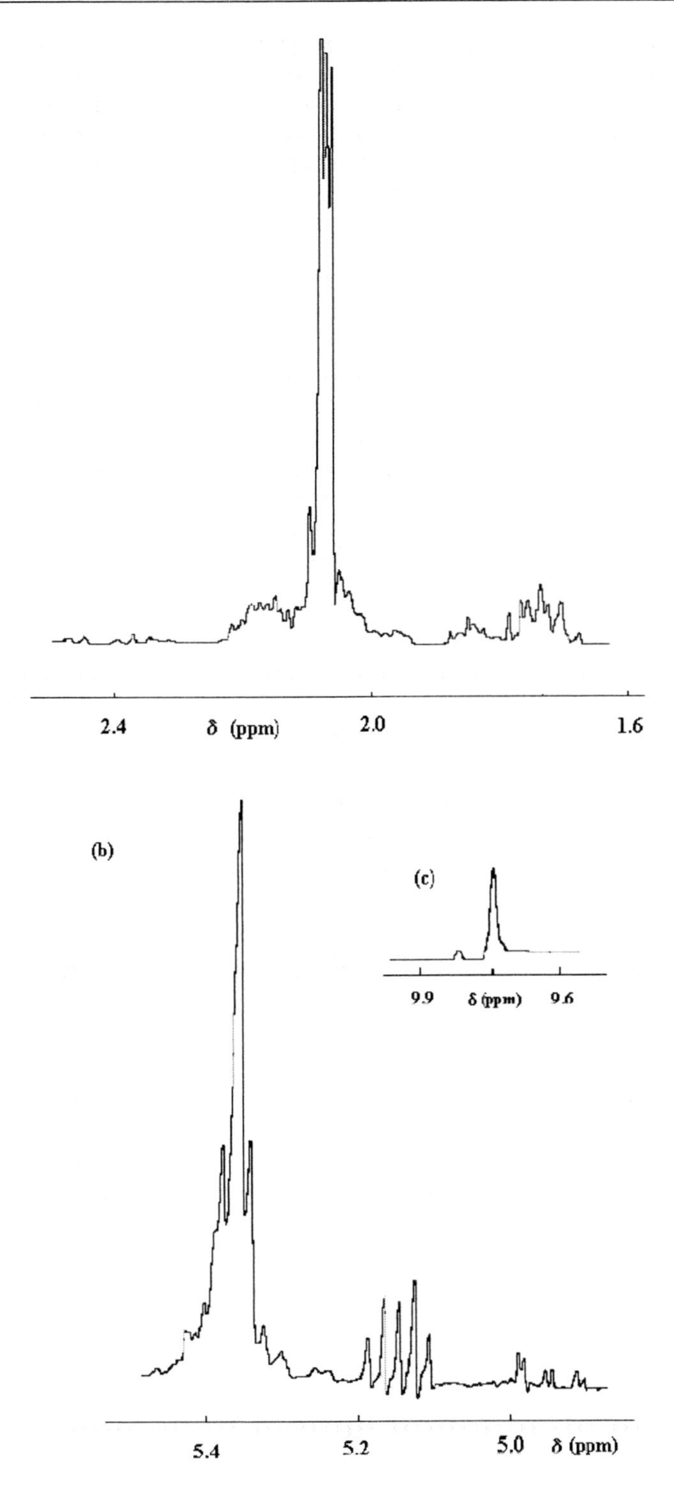

Figure 4. ^{1}H-250 MHz NMR spectra of E-BR solutions (0.89 g / 100 ml CCl_4) ozonized to 18 % conversion of the double bonds (external standard TMS; digital resolution 0.4 Hz, 20°C).

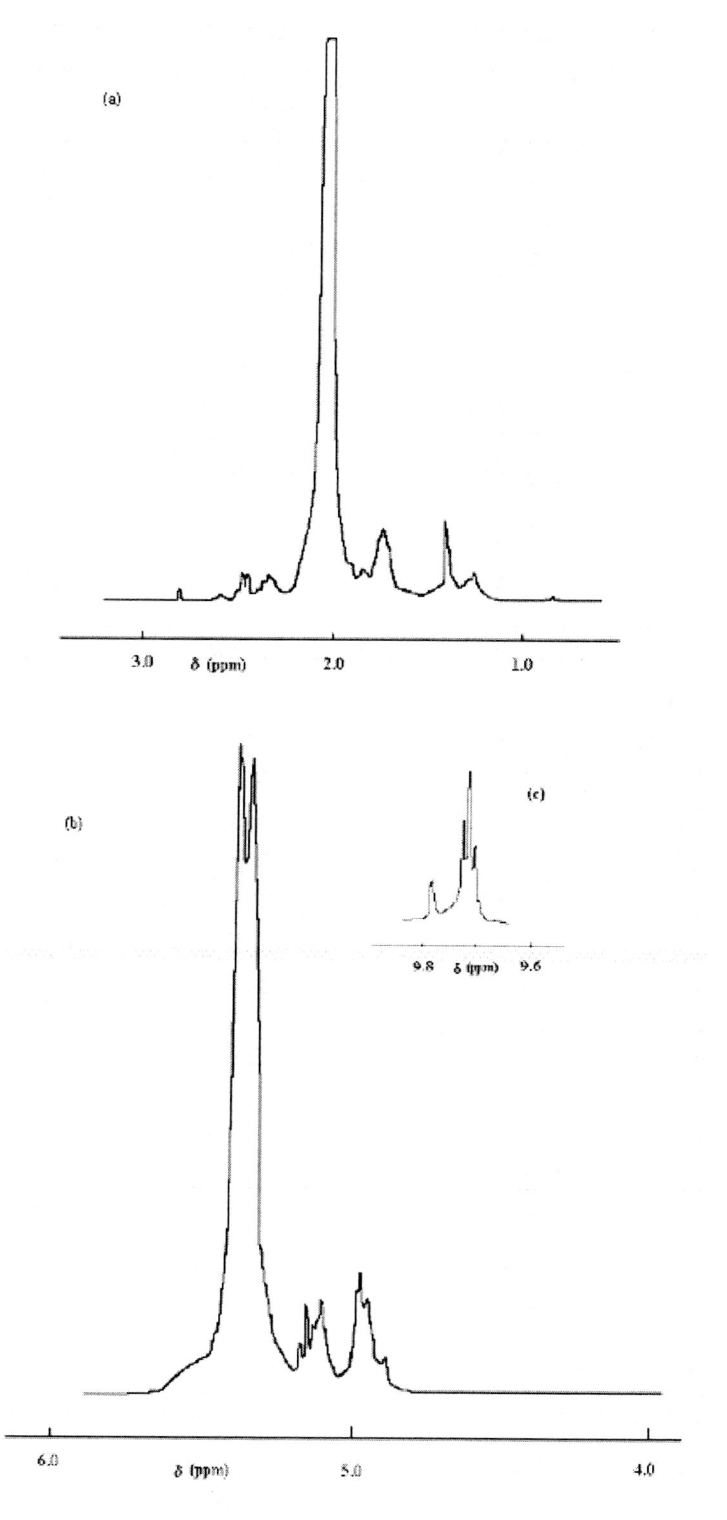

Figure 5. ^{1}H-250 MHz NMR spectra of BR solutions (0.89 g / 100 ml CCl_4) ozonized to 18 % conversion of the double bonds (external standard TMS; digital resolution 0.4 Hz, 20°C).

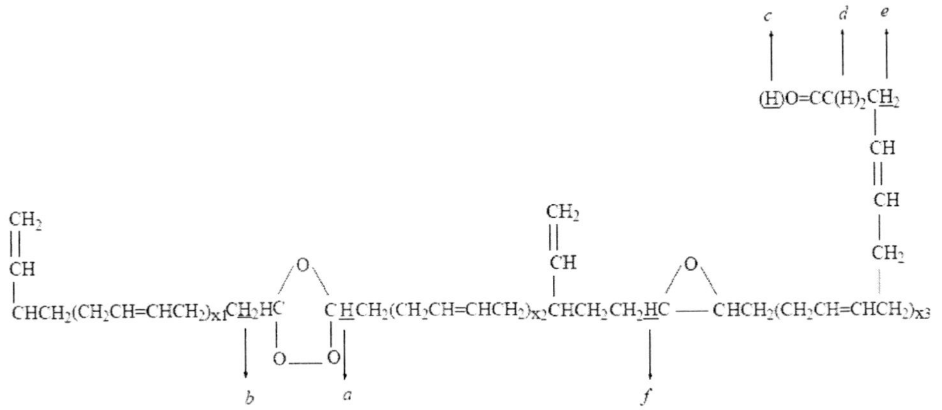

c d e

(H)O=CC(H)$_2$CH$_2$ | CH || CH | CH$_2$

CH$_2$ || CH | CHCH$_2$(CH$_2$CH=CHCH$_2$)$_{x1}$CH$_2$HC(O–O–O)CHCH$_2$(CH$_2$CH=CHCH$_2$)$_{x2}$CHCH$_2$CH$_2$HC — CHCH$_2$(CH$_2$CH=CHCH$_2$)$_{x3}$

b a f

where x1, x2, x3 = 1, 2, 3.......n

Figure 6. Selection of protons with characteristic signals in the ^{1}H-250 MHz NMR spectra of partially ozonized polybutadiene macromolecules.

Table 1. Assignment of the signals in the ^{1}H-NMR spectra of partially ozonized E-BR and BR rubbers.

Assignment of the signals (according Fig. 4)	Chemical shifts (ppm)		Literature
	E-BR	BR	
a	5.10 - 5.20 max 5.12, 5.16	5.05 - 5.18 max 5.10, 5.15	[12, 26]
b	1.67 - 1.79 max 1.72, 1.76	1.66 - 1.80 max 1.73	[12, 26]
c	9.75	9.74	[27]
d	2.42 - 2.54 max 2.47	2.42- 2.54 max 2.50	[12, 26]
e	2.27 - 2.42 max 2.35	2.27 - 2.42 max 2.35	[12, 26]
f		max 2.81	[27]

There exist two isomeric forms of 1,2,4-trioxolanes [2, 10]. Their ratio is a function of the double bond stereochemistry, the steric effect of substituents and the conditions of ozonolysis and it has been studied only for low molecular weight alkenes [10, 27]. The ^{1}H-NMR spectroscopy is the most universal method for determination of the cis/trans ratio of ozonides (in the case of polymers it is practically the only method that can be applied). The measuring is based on the differences in the chemical shifts of the methine protons of t he two isomers: the respective signal of the cis form appears in lower field compared to the trans one [2, 10]. In accordance with the above mentioned, the multiplet on Figure 3 in the area of 5.1-5.18 ppm could be interpreted as a result of partial overlapping of triplets of trans- and cis-ozonides: 5.12 (t, J≈5 Hz, 2H) and 5.16 (t, J≈5 Hz, 2H), respectively. It is interesting to note that the cis/trans ratio of the E-BR 1,2,4-trioxolanes is practically equal to that obtained from cis-3-hexene [10, 25]. The resolution of the respective BR spectrum does not allow consideration in detail of the multiplicity of the signals at 5.10 and 5.15 ppm. In this case the area of the signals is widened, most probably due to the presence of ozonide signals of the 1,2-monomer units [28].

Reaction Mechanism

The modern concepts about the mechanism of the reaction of ozone with C=C double bonds in solution are summarized in Schemes 1 and 2 [1, 2, 27-29]. The initial reaction product, the cis or trans molozonide (MO) is unstable and at temperatures higher than -150 °C for the cis and - 90 °C for the trans-isomer is decomposed to aldehyde and zwitterionic species termed as Criegee's intermediate or carbonyl oxide (hereafter referred to as CI) (Scheme 1, reaction 2) [27, 28]. The basic product of double bonds ozonolysis in non-participating solvents – the ozonides are obtained in reaction of CI with a carbonyl compound. It has been found out that the dominant part of ozonides is formed through interaction between carbonyl oxide and the corresponding carbonyl group, which both originate from the decomposition of one and the same MO, i.e. a solvent cage effect is acting (Scheme 1, reactions 3 and 3') [1, 27]. The yield of the so-called normal ozonides from the simple olefins are usually over 70 % from the total ozonide yield [27].

The cross-ozonide formation, observed with alkenes [1, 2, 10, 11], is possible with polydienes too (Scheme 1, reaction 5). The problem is that the application of ^{1}H-NMR spectroscopy, in the case of polybutadienes, does not allow differentiation between normal, cross and polymeric ozonides. Our experimental conditions (respective values of [C=C], [O_3] and temperature) do not favour the proceeding of reaction 5 [2]. It is obvious that the influence of the polymeric nature of the double bond substituents on the cross-ozonide formation also cannot be positive.

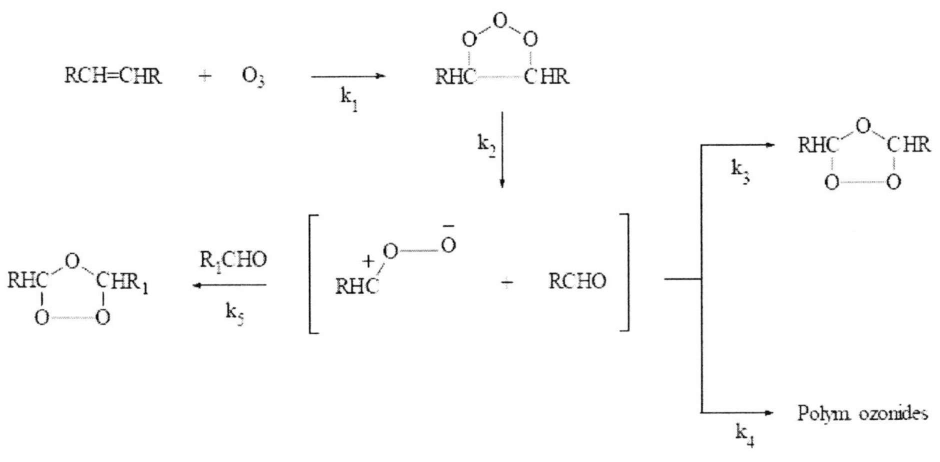

RCH=CHR + O_3 → (k_1) RHC–CHR (molozonide)

k_2

[RHC^{+}–O–O^{-} + RCHO]

k_3 → RHC–O–CHR ozonide

k_4 → Polym. ozonides

R_1CHO, k_5 → RHC–O–CHR$_1$ ozonide

Scheme 1. Formation of ozonides.

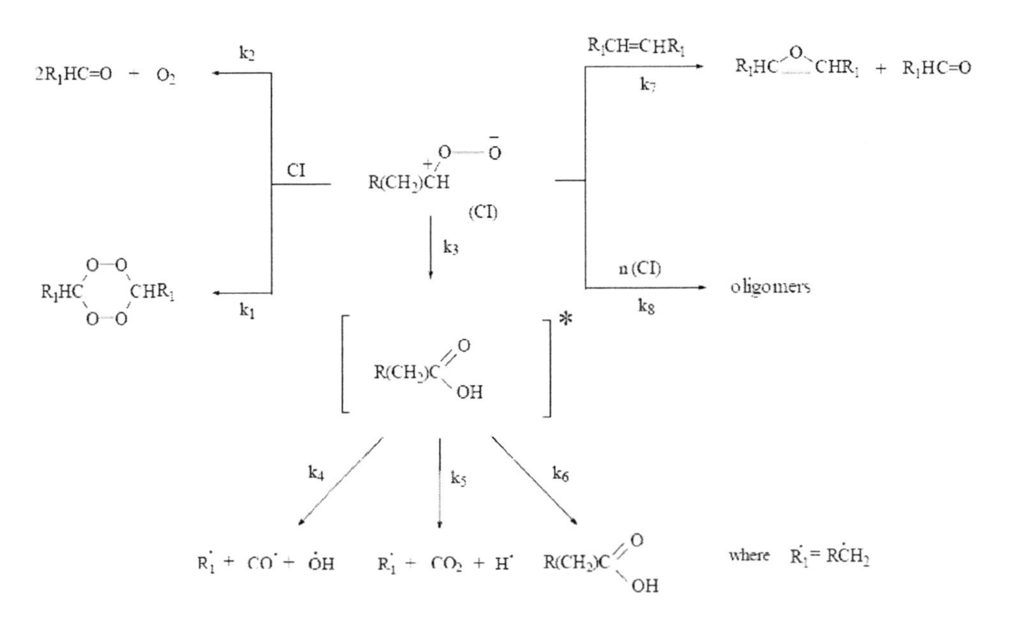

Scheme 2. Alternative routes of conversion of CI intermediates.

As it was already mentioned the ozonide yield of E-BR is considerablyy higher in comparison to that of BR. Upon investigationg of the alkene ozonolysis in solution it has been found out that the yield of normal and cross 1,2,4-trioxolanes depends on such factors as stereometric structure of the double bonds, nature of the substituents and of the solvent, temperature and alkene concentration [2, 10, 25]. There exists a general rule that the cis-alkenes usually give a higher yield of ozonides (total) and of normal 1,2,4-trioxolanes than the trans isomers [2, 10]. For example the maximal ozonide yields of the cis- and trans 3-hexene are respectively 88 and 49 %, and the minimum values are 81 and 22 % [2, 10]. In the case of cis and trans 2-butene the maximum values are 72 % and 36 % and the minimum 48 % and 36 % correspondingly [2, 10]. The high selectivity of the reaction towards 1,2,4-trioxolanes is also typical of the ozonolysis of the terminal double bonds which could be considered as analog of the 1,2- monomer units [28]. The analysis of the reaction selectivity towards the ozonides shows similarity between the 1,4-cis-polybutadiene units and the cis-3-hexene, while in the case of cis-2-butene the results are lower. An attempt was made to evaluate the yield from 1,4-trans units, based on their contents in BR, the data determined by ^{1}H-NMR for E-BR and assuming the ozonide yield of 1,2-units equal of that of 1,4-cis units. We obtained that the selectivity towards ozonides for the 1,4-trans units of polybutadiene is about 50 %.

As it was reported previously at 20-25 % ozone conversion of polybutadienes, especially of those with high contents of 1,4-cis units a gel formation is observed [17, 22]. It has been found out that the crosslinks of polymers containing double C=C bonds in the side chain have peroxide nature [30]. The kinetics of the crosslinking process during ozonolysis of polymers with double bonds has been studied in [23, 31]. Our expectations are that in the case of the polybutadienes, the predominant part of crosslinks are result of formation of polymeric ozonides (Sheme 1, reaction 4), but their share in the overall balance of reacted ozone is very small, of the order of a few percents.

The second main product of the polybutadiene ozonolysis are the aldehydes. According to the mechanism of the ozone reaction with double bonds in solution, the aldehyde groups are being formed when the conversion of CI intermediates is proceeding through routes, which are an alternative to the carbonyl oxide-aldehyde interaction [2, 29]. The non-ozonide routes are illustrated in Scheme 2. In the case when the CI conversion results in formation of carbonyl group, two moles of aldehydes are obtained from one mol of ozone (Scheme 2, reactions 2 and 7). Another route of CI deactivation is its isomerization via hot acid to radicals (Scheme 2, reactions 4, 5). In this case 1 mol of ozone produces 1 mol of aldehyde [3]. Acidic groups were not detected among the reaction products (Scheme 2, reaction 6). The stability and the life time of low molecular CI intermediates with different substituents and their capability to undergo various monomolecular and bimulecular interactions are widely discussed in the literature [28, 29]. The peculiarities of conversion of two intermediates in polyisoprene ozonolysis (one of them is the same as that obtained from polybutadiene) are considered in Ref. [20]. It has been shown that, under our experimental conditions, the dominant 'non-ozonide' route of conversion of the respective carbonyl oxide intermediate is its isomerization via hot acid.

The epoxide groups may be formed as a result of direct interaction between ozone and a double bond, or by a reaction of a double bond with dioxirane, which could be considered as the cyclic form of the carbonyl oxide intermediate [28, 29]. However there is no evidence that the above mentioned reactions or the cyclization of the respective intermediate are taking place with alkenes, which are structural analogues of the polybutadienes, or with the very polybutadienes. The most probable suggestion is that the epoxide groups are products of interaction of the carbonyl oxide with a double bond (Scheme 2, reaction 7). It should be noted that, in contrast to the ozonolysis of cis- and trans-hexene [25], 1,4-cis- and 1,4-trans-polyisoprene [20], in the case of polybutadienes the formation of epoxide groups is observed only when the elastomer macromolecules contain 1,4-trans monomer units. This phenomenon could be explained with the relatively higher stability of the dominant stereochemical form (syn carbonyl oxide) of the respective intermediate that originates from ozonolysis of 1,4-trans units [29].

The basic route of the reaction - the formation of normal ozonides does not lead directly to a decrease in the molecular mass of the elastomer macromolecules, because the respective 1,2,4-trioxolanes are relatively stable at ambient temperature [31, 32]. The most favourable conditions for ozone degradation emerge when the cage interaction (Scheme 1, reaction 3) does not proceed. Therefore the higher the ozonide yield the lower the intensity of ozone degradation of the polybutadienes and vice versa. As it was already determined the ozonide yields for the 1,4-cis- and 1,2- monomer units are of the order of 83-90 %, whereas that for the 1,4-trans units is about 50 %. The amount of aldehyde groups is usually used for evaluation of the intensity and efficiency (number of chain scissions per molecule of reacted ozone) of ozone degradation of elastomers. In this case it should be taken into account that the dominant route of degradation leads to the formation of 1 mol of aldehyde from 1 mol of ozone.

CONCLUSIONS

The reaction of ozone with 1,4-cis-polybutadiene and polybutadiene containing 1,4-cis-; 1,4-trans- and 1,2-monomeric units was investigated in CCl_4 solution. The aldehyde:ozonide ratio was 11:89 and 27:73 for E-BR and BR, respectively. In addition, epoxide groups were detected, only in the case of BR and their yield was about 10 % of that of the aldehydes. Based on the BR ozonolysis, it was established that the ozonide yield from 1,4-trans units is about 50 %.

The non-ozonide routes of deactivation of the Crigee's intermediates were discussed. A supposition is substantiated that monomolecular decomposition to radicals via isomerization to hot acid is the dominant route for CI conversion. Taking into account the aldehyde yields, an evaluation was made of the efficiency of ozone degradation of the two polybutadienes, according to which the respective value of BR is considerably higher than that of E-BR.

REFERENCES

[1] R. Crigee, *Angew. Chem. Int. Ed. Eng.* 1975, 14, 745.

[2] P. S. Bailey, *Ozonation in organic chemistry*, Vols 1, 2, Academic Press, New York 1978, 1982.

[3] S. D. Razumovskii; G. E. Zaikov, *Ozone and its reactions with organic compounds*, Elsevier, Amsterdam 1984.

[4] S. Rakovsky; G. Zaikov; *Kinetics and mechanism of ozone reactions with organic and polymeric compounds in liquid phase*, Nova Science Publishers Inc., New York, 1998.

[5] D. G. Williamson, R. J. Cvetanovic, *J. Am. Chem. Soc.* 1968, 90, 4248.

[6] W. Pritzkow, G. Schappe, *J. Prakt. Chem.* 1969, 311, 689.

[7] D. Cremer, *J. Am. Chem. Soc.* 1981, 103, 3619.

[8] Ch. R. Greene, R. Atkinson, *Int. J. Chem. Kinet.* 1992, 24, 803.

[9] S. D. Razumovskii, *Khimicheskaja Physika*, 2000, 19 (7), 58.

[10] R.W. Murray, R.D.Youssefyeh, P. R. Story. *J. Am. Chem. Soc.* 1967, 89, 2429.

[11] J-I. Choe, M. Sprinivasan, R. L. Kuczkowski, *J. Am. Chem. Soc.* 1983, 105, 4703.

[12] M. J. Hackathorn, M. J. Brock, *Rubber Chem. Technol.* 1972, 45, 1295.

[13] I. Spasskova, *Vestn. Leningr. Univ., Khimia*, 1973, No 10, 140.

[14] G. G. Egorova, V. S. Shagov, "Ozonolysis in the Chemistry of Unsaturated Polymers", in: *Synthesis and Chemical Transformation of Polymers*, Izd. LGU, Leningrad, 1986. (In Russian).

[15] Yu. S. Zuev, T. G. Degtereva, *Elastomer stability under operating conditions*, Khimia, Moscow, 1986 (In Russian).

[16] D. Bruck, *Kautch.+Gummi Kunststoffe*, 1989, 42, 760.

[17] M. P. Anachkov, S. K. Rakovsky, D. M. Shopov, S. D. Razumovskii, A. A. Kefely, G. E. Zaikov, *Poly. Deg. and Stab.* 1985, 10, 25.

[18] G. Ivan, M. Giurginka, *Poly. Deg. and Stab.* 1998, 62, 441.

[19] F. Cataldo, Polym. Degrad. Stab. 2001, *73*, 511.

[20] M. P. Anachkov, S. K. Rakovsky, R. V. Stefanova, *Poly. Deg. and Stab.* 2000, 67, 355.

[21] M. P. Anachkov, Unpublished results.

[22] S. D. Razumovskii, M. P. Anachkov, A. A. Kefely, G. E. Zaikov, *Visokomolek. Soed. Ser. B*, 1982, 24, 94.
[23] M. P. Anachkov, S. K. Rakovsky, R. V. Stefanova, D. M Shopov, *Poly. Deg. and Stab.* 1987, 19, 293.
[24] M. P. Anachkov, S. K. Rakovsky, R. V. Stefanova, A. K. Stoyanov, *Poly. Deg. and Stab.* 1993, 41, 185.
[25] R. W. Murray, W. Kong, S. N. Rajadhyaksha, *J. Org. Chem.* 58, 1993, 315.
[26] *The Sadtler Handbook of Proton NMR Spectra*, Sadtler, Philadelphia, PA, 1978.
[27] R. L. Kuczkowski, *Chem. Soc. Reviews*, 1992, 79.
[28] K. J. McCullough, M. Nojima, "Peroxides from Ozonization", in: *Organic Peroxides*, Ando W., Ed., John Wiley and Sons Ltd, 1992.
[29] W. H. Bunnelle, *Chem. Rev.* 1991, 91, 335.
[30] M. P. Anachkov, S. K. Rakovsky, D. M. Shopov, S. D. Razumovskii, A. A. Kefely, G. E. Zaikov, *Poly. Deg. and Stab.* 1986, 14, 189.
[31] M. P. Anachkov, S. K. Rakovsky, A. K. Stoyanov, *J. Appl. Polym. Sci.*, 1996, 61, 585.
[32] M. P. Anachkov, S. K. Rakovsky, *Bulg. Chem. Comm.* 2002, 34 (3/4), 486.

In: Reactions and Properties of Monomers and Polymers ISBN: 1-60021-415-0
Editors: A. D'Amore and G. Zaikov, pp. 165-198

Chapter 9

REACTIONS AND STRUCTURE FORMATION OF POLYMERS IN FRACTAL SPACES

G. V. Kozlov and G. E. Zaikov*
Agrarian State University, Voroshilov st., 25, Dniepropetrovsk, 49 027, Ukraine
*Institute of Biochemical Physics of Russian Academy of Sciences
4, Kosygin st., Moscow 119991, Russian Federation

ABSTRACT

The examples described in present paper show strong effect of a space type on reactions and structure formation for polymeric materials. A purposeful change of a space type will allow vary sharply to reaction rate depending on what is this reaction wishing (synthesis) or not (degradation). The offered technique also allows to simulate and predict a structure of composites polymer matrix depending on conditions of an initial filler aggregation.

Keywords: Chemical reactions, structure formation, Euclidean and fractal spaces, thermooxidative degradation, polycondensation, polymeric composites.

1. INTRODUCTION

Theoretically a problem of structure change of this or that object formated not in Euclidean but in fractal space is studied in a number of papers [1-3]. However there is no practical application of this important aspect at present. In present paper several applications of structures formation theory to such practical problems as thermooxidative degradation, synthesis polymers and polymeric composites can be considered.

2. Thermooxidative Degradation of Solid Polymers

As it known [4], the structure of solid polymers is a fractal in interval of linear scales from several Ångströms up to several tens Ångströms. Right in this interval sizes of free volume voids get, through which oxidant diffusion (specially gaseous, for example, oxygen) realizes to reactive centers of polymer macromolecules the laws, describing the transport on fractal objects, principally differ from corresponding laws for Euclidean objects [5]. In connection with these two examples are cited [6, 7]. If a trajectory of a molecule of gaseous diffusant will be considered as random walk with rms displacement ξ, the number of such displacements N_W is defined by a relationship [8]:

$$N_W \sim \xi^{d_W}, \tag{1}$$

where d_W is fractal dimension of walk.

For Euclidean spaces $d_W = 2$ is independent from their dimension, whereas for fractals dimension d_W is substantially larger 2 [8]. Then, a number of displacements N_W at equal ξ for fractals will be substantially larger than for Euclidean objects. In this case substantially larger displacement for achievement of reactive center of a polymer-macromolecule shoud be made that shouls lead to the changes a thermooxidative degradation kinetics.

The number of sites visited by random walks (achieved contact centers) $\langle S \rangle$ is defined as follows [8]:

$$\langle S \rangle \sim N_W^{d_s/2}, \tag{2}$$

where d_s is spectral dimension characterizing a connectivity degree of the object [9].

For Euclidean objects it is with dimension $d = 3$ $d_s = d = 3$ whereas for fractal objects $d_s < 2$ [8]. From a relationship (2) follows, that the number of sites visited by random walk will be substantially less in fractal space than in Euclidean. This means, that in case of a thermooxidative degradation in the first case molecule of diffusant can contact with a less number of reactive sites of a polymer macromolecule than in the second case. Therefore it should be expected that the thermooxidative degradation rate for fractal solid polymers will be substantially less than supposed for Euclidean solid bodies.

In paper [10] a theoretical treatment of trapping of random walk (oxygen molecule) by traps (that can be considered as particles of nonchain inhibitor – high – dispersed mixture Fe/FeO(Z) [6, 11] on fractal object, is considered. The average survival probability of the random walks $\Phi(t)$ after time t is supposed to be given by [10]:

$$\Phi(t) \sim \exp\left(-c_1 p^{\beta} t^{\gamma}\right), \tag{3}$$

where c_1 is constant, p is the fraction of traps and exponents β and γ are defined as follows [10]:

$$\gamma = \frac{d_s}{2 + d_s}, \tag{4}$$

$$\beta = 1 - \gamma. \tag{5}$$

Following a model on nonchain inhibition [12] authors [6] assumed that the stabilizing action of Z is preserved until survival probability of oxygen molecules $\Phi(t)$ is very small, i.e., until practically all oxygen is trapped by traps (particles Z with concentration p). Such treatment is agreed with experimental data obtained at aging of high density polyethylene (HDPE) stabilized by nonchain inhibitor Z (compositions HDPE + Z) [12]. These results show constancy of studied property (impact toughness A_p, melt flow index MFI, tangent of dielectric loss angle tgδ) during some period of aging (called induction period t_{in}) after that a rather sharp change of this property begins [6, 11-13]. Therefore, supposing for $\Phi(t)$ small enough (close to zero) value, for example 0,03, the value t from a relationship (3) can be estimated during which the condition $\Phi(t) = 0{,}03$ (t_{in}) is preserved. Technically the solution of this task requires the determination of the spectral dimension d_s. After that with the use of a relationship (2) the value $\langle S \rangle$ can be determined in the assumption that this value is reciprocal to induction period duration t_{in} [6, 7].

Let's consider the question of d_s estimation. Such estimation can be easily made if the fractal dimension of polymer structure d_f is known [8]. However, heat aging processes proceed at high enough temperature (specially if the main role in these processes plays a thermooxidative degradation [14]), which are close or more than glass transition (melting) temperature of the polymer. As Kopelman showed [15], in such conditions the behavior of polymer is defined by effective spectral dimension d'_s. This corresponds to theorem on subordination [15]:

$$d'_s = \beta_p d_s, \tag{6}$$

where β_p is the parameter characterizing distribution of relaxation times.

As parameter β_p is rather difficult to define experimentally [15], authors [6] determined d'_s from experimental data according to the change of some of the mentioned above properties at aging. If the change rate of this property k is obeyed to time dependence [15]: $k \sim t^{-h}$ (h is heterogeneity exponent), this means fractal-like behaviour during aging process and value d'_s is determined as follows [15]:

$$d'_s = 2\,(1 - h). \tag{7}$$

As follows from the plots of properties change for compositions HDPE + Z (MFI and A_p [13], tgδ [12]) a very fast decay for all compositions HDPE + Z is observed except the composition with content of Z 0,05 mas. %. This means, that the value d'_s for the latter from the mentioned above compositions is substantially less than for the remaining ones. The quantitative estimations gave such results: $d'_s \approx 1{,}2$ for composition with content of Z 0,05 mas. % and $d'_s \approx 1{,}8$ for remaining compositions HDPE + Z [6].

As the comparison of experimental data at heat aging of compositions HDPE + Z lower (80 °C) and higher (190 °C) melting temperature showed values t_{in} in the first case have order of about 10^3 hours and in the second case – several hours, i.e., difference of about three orders of magnitude is observed. This difference is easily explained within the model offered in paper [6] if to allow for that the heat aging processes are proceeding in the first case in fractal space, and in the second case – in Euclidean (or close to Euclidean) space [4]. As it is known [10], for Euclidean objects $d_s = d$ (where d is dimension of Euclidean space), i.e., in considered case $d_s = d = 3$. The combination of relationships (2) and (3) allows to obtain the following relationship for estimation $\langle S \rangle$ [6]:

$$\langle S \rangle \sim \frac{C_1^{d_s/2\beta}}{p^{\gamma d_s/2\beta}}. \tag{8}$$

At C_1 = const and the mentioned above values d_s in case of composition HDPE + Z with content of Z 1,0 mas. % in paper [6] is obtained $\langle S \rangle \sim 10^2$ in fractal space and $\langle S \rangle \sim 10^5$ – in Euclidean space. The time t_{in} is made about 10^{-2} and 10^{-5} relative units in the first and second cases correspondingly, This simple estimation gives the same difference in three orders of magnitude as experimental data [12, 13].

As it was shown in paper [16], the dependence of the relative impact toughness A_p/A_p^0 (where $\overline{A}_p$ and A_p^0 are impact toughness of specimens after the arbitrary duration of heat aging and impact toughness of initial specimen, accordingly) on the duration of aging t_{ag} (the aging temperature 393 K) for polybutyleneterephthalate (PBT) breaks up two regions. In first region where the fast decay of A_p/A_p^0 in the course of aging is observed, pseudo-monomolecular reaction is realized, and in the second one, where the decrease of A_p/A_p^0 with t_{ag} occurs much more slowly, the bimolecular reaction is realized. In means that in the first region the thermooxidative degradation of PBT is caused by the reaction of the intermediate products of degradation with a polymer macromolecule, and in the second one – the reaction of oxygen with a macromolecule [16]. In paper [17] the structural aspects of these processes were examined and it was found out that the pseudo-monomolecular reaction proceeds in the loose-packed (devitrificated at the temperature of aging) areas of an amorphous phase of PBT, and the bimolecular reaction – in the dense-packed areas (areas of the local order (clusters) of an amorphous phase and crystallites) of polymer. It is possible to obtain the additional information about the processes occuring in the course of thermooxidative degradation of PBT by the examination of its kinetics within the framework of the fractal kinetics of chemical reactions [15, 18] and this will be show below.

In Figure 1 the dependence of average weight molecular weight $\bar{M}_W$ on heat aging duration t_{ag} for PBT is given, from which the fast decay of $\bar{M}_W$ within approximately first ten days follows, and then this decay sharply slows down. As mentioned above, the thermooxidative degradation rate k can be described by the relationship $k \sim t_{ag}^{-h}$. In case of the course of reactions in fractal mediums $h > 0$, and in case of Euclidean medium (classical behaviour) $h = 0$ and then $k =$ const [15, 18]. The value k can be calculated as follows [19]:

$$k = \frac{\left(\bar{M}_W\right)_{i+1} - \left(\bar{M}_W\right)_i}{\left(t_{ag}\right)_{i+1} - \left(t_{ag}\right)_i}, \tag{9}$$

where $\left(\bar{M}_W\right)_{i+1}$ and $\left(\bar{M}_W\right)_i$ are molecular weights of polymer at arbitrary aging durations $\left(t_{ag}\right)_{i+1}$ and $\left(t_{ag}\right)_i$, accordingly.

The dependence $k(t_{ag})$ in double log-log coordinates allows to determine the value of heterogeneity exponent h. In Figure 2 and Figure 3 such dependences for regions of the pseudomonomolecular and bimolecular reactions of thermooxidative degradation of PBT are given, accordingly. In the first case (Figure 2) the reaction rate k does not depend on t_{ag}, i.e., in this case we gain the classical behaviour (the course of reaction in an Euclidean space), where $k =$ const and $h = 0$. As in case of polymers, an Euclidean object can be only the devitrificated (rubber-like) polymer [4], the obtained data confirm the conclusion of paper [17] about proceeding of the pseudo-monomolecular reaction in the process of PBT thermooxidative degradation in the devitrificated areas of an amorphous phase. In Figure 4 for comparison the dependence $k(t_{ag})$ for heat aging of PBT over the melting temperature $T_m \approx$ 500 K [20], i.e., for its melt is shown. In this case, the whole specimen is an Euclidean object [4] and, as it was necessary expected, for it we also obtain $k =$ const, $h = 0$ (classical behaviour). Thus, the principal distinctions of the proceeding of the pseudo-monomolecular reactions for the devitrificated areas of an amorphous phase of solid PBT (at $T_g < T_{ag} < T_m$, where T_g and T_{ag} are glass transition and heat aging temperatures, accordingly) and completely devitrificated PBT do not exist as the mentioned structural areas are the Euclidean objects. However, the pseudo-monomolecular reaction rate in melt is more approximately in 20 times than for solid polymer.

In Figure 3 the dependence $k(t_{ag})$ in double log-log coordinates for a region of realization of the bimolecular reaction of thermooxidative degradation of PBT (the second region of the dependence $\bar{M}_W(t_{ag})$ in Figure 1) is shown. As it is possible to see the decay k in this case is observed at the t_{ag} increase, that is the proof of the proceeding of the bimolecular reaction in fractal space [15, 18]. The linearity of the dependence ln k(ln t_{ag}) allows to calculate the heterogeneity exponent h from its slope, which is equal to ~ 1,0. As mentioned above, the effective spectral dimension d'_s can be determined according to the equation (7). At $h = 1$, $d'_s = 0$, i.e., the bimolecular reaction proceeds in zero-dimensional space or, on the definition of the authors [18], in miniclusters. Such miniclusters for semicrystalline polymers are dense-packed areas (clusters and crystallites) [21], and it also confirm the conclusion of paper [17] about the proceeding of the bimolecular reaction in dense-packed areas of PBT. It is possible to see that the pseudo-monomolecular and the bimolecular reactions represent the extreme cases of the thermooxidative degradation kinetics, which are relevant to boundary values h = 0 and h = 1, accordingly [15, 18]. On this basis can be assumed that the values k, obtained for the mentioned reactions, give the limiting (maximal and minimal, accordingly) thermooxidative degradation rates. In Figure 3 the dependence ln k (ln t_{ag}) for polyarylate on the basis of phenolphthalein (PAr) is also given [2]. As follows from the data of Figure 3, the mentioned dependence for PAr is linear and from its slope it is possible to determine h = 0,385. It means, that the thermooxidative degradation of PAr proceeds in fractal space, as it was expected, as heat aging of PAr was carried out at $T_{ag} < T_g$ [22]. The value h for PAr has the intermediate magnitude in an interval of the possible magnitudes $0 \leq h \leq 1$ and this circumstance assumes the simultaneous proceeding of thermooxidative degradation both in loose-packed, and in dense-packed (clusters) of PAr structural regions. The rather small value h for PAr allows to assume that the main destructive processes proceed in the loose-packed areas of this polymer structure [23].

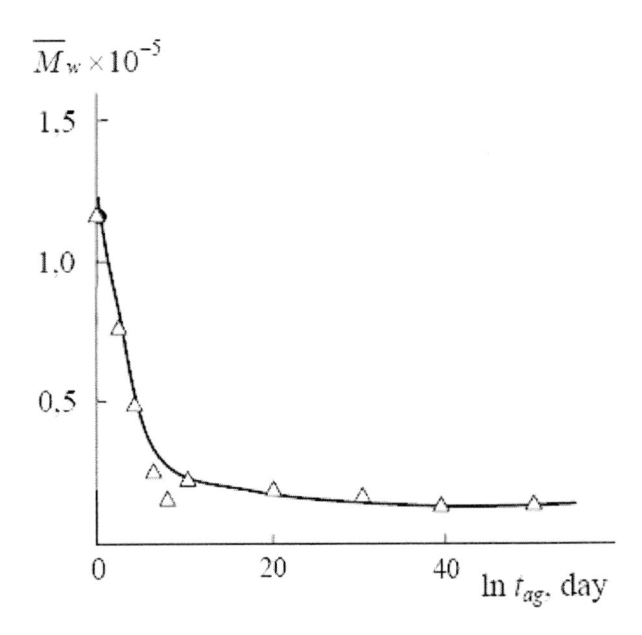

Figure 1. The dependence of molecular weight $\overline{M}_W$ on aging duration t_{ag} for PBT [23].

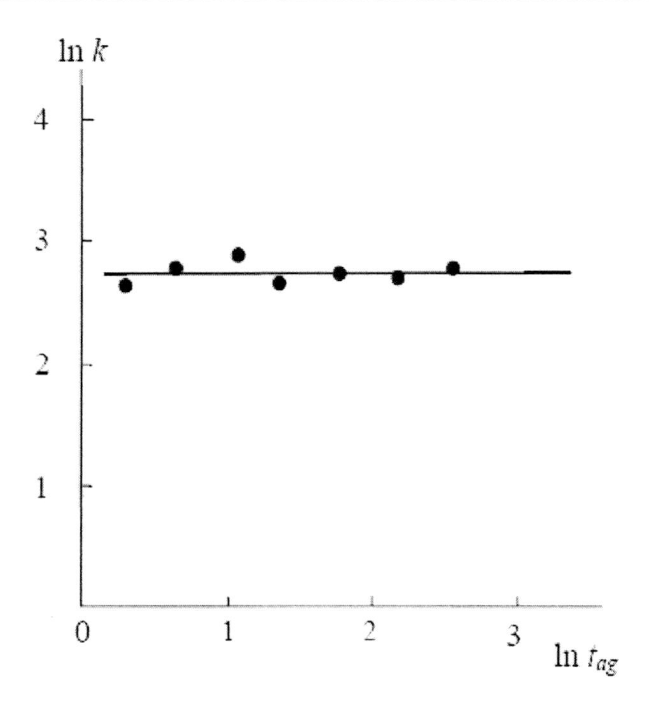

Figure 2. The dependence of pseudo-monomolecular reaction rate k on heat aging duration t_{ag} in double log-log coordinates for solid PBT [23].

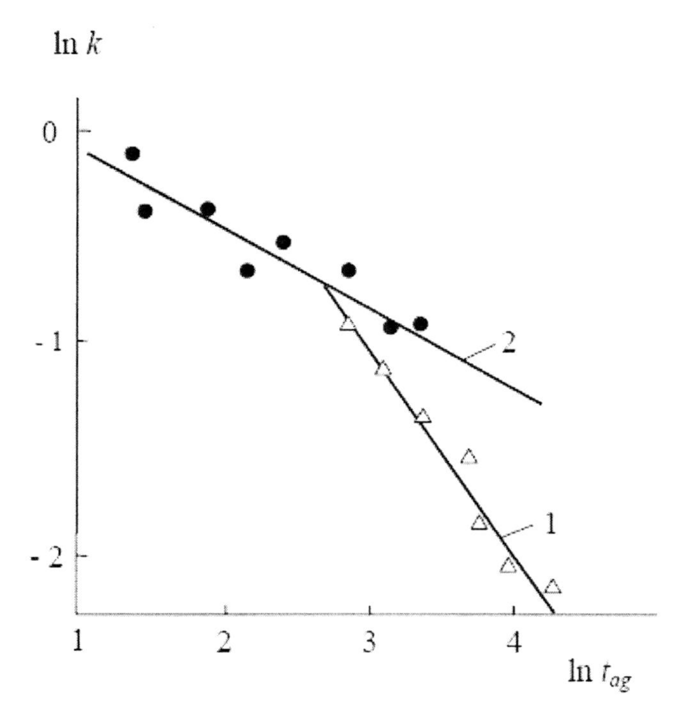

Figure 3. The dependence of reaction rate k on heat aging duration t_{ag} in double log-log coordinates for the bimolecular reaction of PBT (1) and thermooxidative degradation of PAr (2) [23].

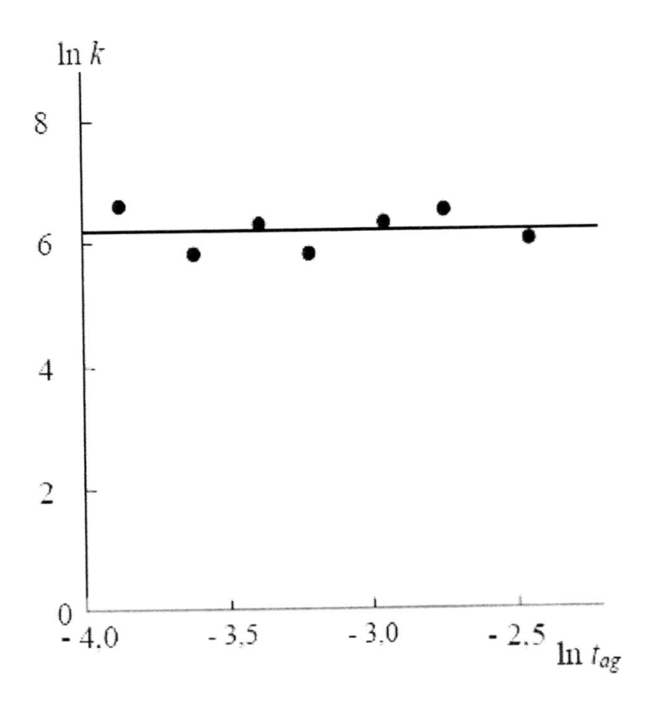

Figure 4. The dependence of pseudo-monomolecular reaction rate k on heat aging duration t_{ag} in double log-log coordinates for a melt of PBT [23].

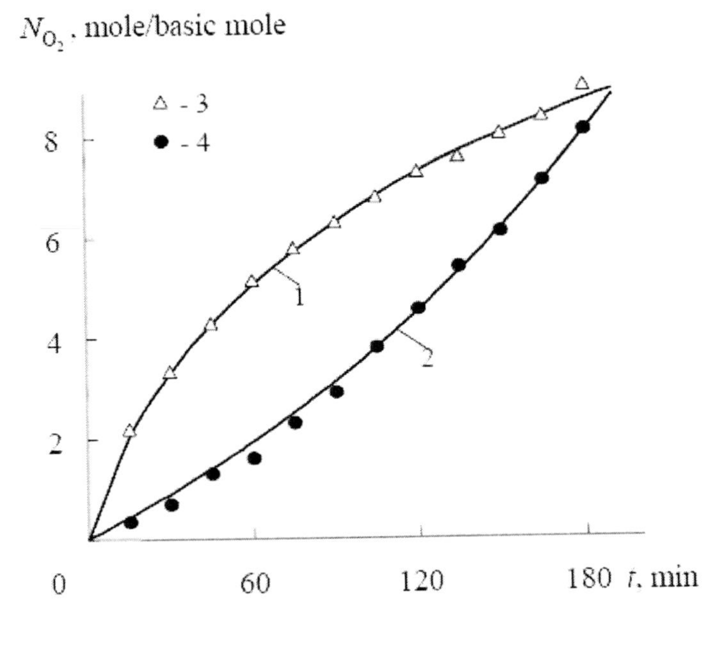

Figure 5. The experimental kinetic curves amount of consumed oxygen-time $N_{O_2}(t)$ for PAASO-1 (1) and PAASO-2 (2) at $T = 623$ K. Calculation of curves $N_{O_2}(t)$ according to the equations (24) (3) and (25) (4) [32].

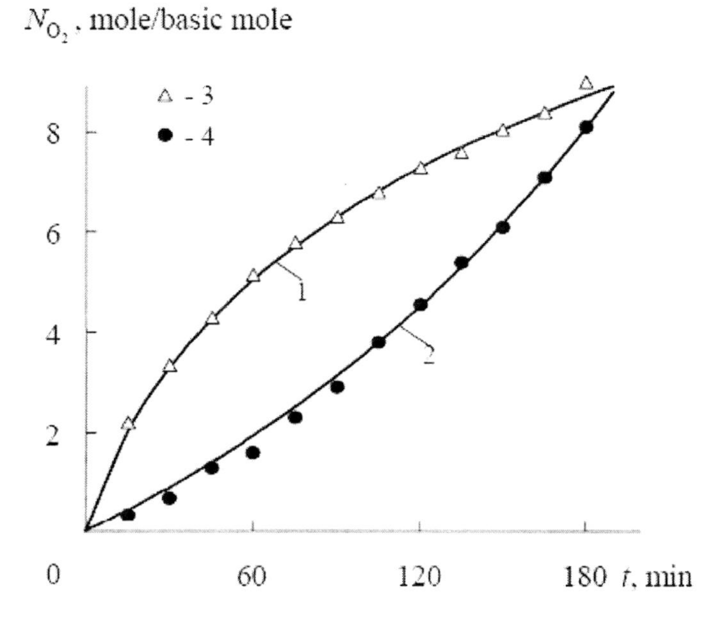

Figure 6. The dependences of consumed oxygen amount on time t in double log-log coordinates for PAASO-1 (1) and PAASO-2 (2) at T = 623 K [32].

Thus, the cited above results have shown that the examination of the kinetics of PBT thermooxidative degradation within the framework of a fractal model [15, 18] allows the exact identification of structural regions, in which the pseudo-monomolecular and bimolecular reactions are realized. The main reason for the realization of the mentioned reactions is the type of space: the pseudo-monomolecular reaction proceeds in an Euclidean space, and the bimolecular reaction proceeds in a fractal space [23].

3. Thermooxidative Degradation at Aging of Polymeric Melts

Now it is well known [24], that there are two main types of kinetic curves amount consumed oxygen-time $N_{O_2}(t)$ in processes of polymers thermooxidative degradation: on autoaccelerated type curves with the legibly expressed induction period and an autodecelerated type curves. It is assumed [24], that the type of kinetic curve $N_{O_2}(t)$ is defined by chemical constitution of polymer and testing temperature. However, in such treatment the structural aspects, which as shown in [25, 26], play a determining role in realization, of this or that type of curves $N_{O_2}(t)$, are left out.

In the last 20 years large attention was given to physical aspects of reaction proceeding of type [27 – 31]:

$$A + A \rightarrow 0, \tag{10}$$

$$A + B \rightarrow 0, \tag{11}$$

where A u B reactions, 0 is inert product.

It is shown, that these reactions can be described by a power function of a type:

$$\rho_A \sim t^x, \tag{12}$$

where ρ_A is an amount of reactant A, reacted during time t, x is exponent.

The value of an exponent x depends on a number of the factors: the space, in which the chemical reaction proceeds (Euclidean or fractal [29]), dimensions characterizing these spaces, and the type of reaction (mono- or bimolecular reaction). Therefore, studing oxidation reactions within the framework of the concepts [27 – 31], it is possible to obtain more full representation about conditions of their proceeding. This problem will be studied in paper [32] on the example of melts of two heterochain polyethers: polyarylate (PAr) and polyarylatearylenesulphonoxide (PAASO). Obtained by two different methods of polycondensation PAASO (table 1) is a convenient object for such studies – at an identical chemical constitution and testing temperature these block-copolymers have discovered curves $N_{O_2}(t)$ of different types [33], that allows to explain this transition by only structural factors [34].

In Figure 5 two kinetic curves $N_{O_2}(t)$ for PAASO-1 and PAASO-2 (table 1) at the temperature of tests T = 623 K are shown. As it is possible to see, these curves fall into different types: for PAASO-1 it is a curve of an autoaccelerated type, for PAASO-2 autodecelerated one. As we consider oxidation of the same polymer at the same testing temperature, the conclusion was drawn, that the change of a curve type is due by structural differences of PAASO-1 and PAASO-2 melts [34].

Let's consider the question of the quantitative description of polymers melts structure. As it is known [35], the temperature of the so-called transition "the liquid 1 – the liquid 2" T_{ll} can be estimated as follows:

$$T_{ll} \approx (1{,}20 \pm 0{,}05)\, T_g. \tag{13}$$

From the equation (13) and data of table 1 the condition follows: $T > T_{ll}$. At T_{ll} there is a transition of a polymeric melt from "liquid with the fixed structure (where the residual structural ordering is observed [35]) to the true liquid state or "structureless liquid" [36]. However, the term "absence of melt structure" at $T > T_{ll}$ concerns the absence of supermolecular structure, but the absence of supermolecular coil in a melt remains the important structural factor (essentially, unique at $T > T_{ll}$) [25, 26].

Table 1. Methods of polycondensation, notation and basic properties of heterochain polyethers [33].

Polymer	The polycondensation method	Notation	T_g, K	$\bar{M}_W \times 10^{-3}$
PAASO	Low-temperature	PAASO-1	472	76
	Emulsive	PAASO-2	489	58
PAr	Low-temperature	PAr	471	76

Most precisely structure of a macromolecular coil, which is a fractal object [37], can be described with the help of its fractal (Hausdorff) dimension Δ_f, describing distribution of a coil elements in space [37].

The number of the reacted molecules of species A ρ_A should in this case obey to a scaling relationship [29]:

$$\rho_A \sim \xi_m^{-\Delta_f}, \tag{14}$$

where ξ_m is characteristic size of a macromolecular coil of a volume V [29]:

$$V \sim \xi_m^{\Delta_f}. \tag{15}$$

The particles of low molecular weight species (molecules of oxidant, radicals etc.) move on the fractal by random walk with dimension d_W, that allows to write [29]:

$$\xi_m \sim t^{1/d_W}. \tag{16}$$

Finally for reaction such as (10) we shall write:

$$\rho_A = N_{O_2} \sim t^{\Delta_f/\Delta_W}. \tag{17}$$

For Euclidean spaces with dimension $\Delta_f = d$ and $d_W = 2$ [29] we shall obtain:

$$\rho_A = N_{O_2} \sim t^{d/2}. \tag{18}$$

For reactions such as (11) relationship (18) changes as follows [29]:

$$\rho_A = N_{O_2} \sim t^{d/4}. \tag{19}$$

As it is known [19], the exponent in relationship (17) can be written as follows:

$$d_s = \frac{2\Delta_f}{d_W}. \tag{20}$$

Further from combination of relationships (17) and (20) it is possible to obtain analogues of relationship (18) and (19) for a case of oxidation reaction proceeding in fractal space. For reactions such as (10):

$$\rho_A = N_{O_2} \sim t^{d_s/2} \tag{21}$$

and for reactions such as (11):

$$\rho_A = N_{O_2} \sim t^{d_s/4} \tag{22}$$

According to the known Alexander and Orbach statement about a superuniversality (independence on *d*) of dimension $d_s = 4/3$ authors [29] have obtained value of an exponent in the relationship (21) equal to 2/3 and in the relationship (22) equal to 1/3. These results were obtained at computer simulation of reactions (10(and (11). However, as mentioned above, in real conditions it is necessary to use not value d_s but value of effective spectral dimension d'_s ($d'_s \le d_s$) defined according to the equation (6).

Now it is possible to plot relationship $N_{O_2}(t)$ in double log-log coordinates and to determine an exponent *x* in relationship (12). In Figure 6 such relationship are shown for curves $N_{O_2}(t)$, shown in Fig. 5, i.e., autoaccelerated and autodecelerated types. As it is possible to see, in both cases the obtained correlations are linear, i.e., are described by the equations such as relationship (12). For PAASO-2 (autoaccelerated oxidation) the value x = 1,5. It means, that in this case under conditions $d = 3$ and $d'_s \le 2$ one version is possible only: this reaction is described by relationship (18). Therefore, the reaction corresponding to autoaccelerated type of curves $N_{O_2}(t)$, represents reaction such as (10) in Euclidean space with d = 3. It is most probable, that the particles of species A are low molecular weight mobile radicals [38]. For PAASO-1 (autodecelerated oxidation) the value x = 0,5 and for six probes of PAr and PAASO in range T = 623 ÷ 723 K the value x is changed within the limits 0,17 ÷ 0,50. Such values x eliminate the possibility of these reactions proceeding in an Euclidean space, where for d = 3 minimal value x is equal to 0,75 (see relationship (19)). For the same reasons it is necessary to eliminate the possibility of reaction proceeding in the fractal spaces with the dimension ds usage, as for them x ≥ 1/3 (see relationships (21) and (22)). For explanation of the obtained range of values x in autodecelerated oxidation case authors [34] use a method of theoretical calculation of dimension d'_s [39], to what the following formula is applied:

$$d'_s = \frac{2(2\Delta_f - d)}{d+2}. \tag{23}$$

At values $\Delta_f \approx 1,9 \div 2,8$ [33] we shall receive $d'_s = 0,32 \div 1,04$. The comparison of these values d'_s and mentioned above values *x* demonstrates, that between them there is a relation $x = d'_s/2$ in error limits of determination of these parameters. Thus, the kinetic curves of an autodecelerated type represent reaction of type (10), proceeding in fractal space and requiring for its description effective spectral dimension d'_s [34]. Let's mark, that the transition from autoaccelerated to autodecelerated oxidation mode happens without change of reaction type, and is unique by virtue of space type change-from Euclidean to fractal. As for Euclidean space $d_s = \Delta_f = d$ [5], in means space connectivity change, characterized by change d_s (for Euclidean space $d_s = 3$, for fractal space $d'_s = 0{,}32 \div 1{,}04$). Let's mark, that at invariable reaction type the amount of free radicals for an autoaccelerated oxidation mode of PAASO three times exceeds this amount for an autodecelerated mode [40].

Theoretically calculated according to the equation (23) values $d'_s\left(d'^{\mathrm{T}}_s\right)$ under condition $d = 3$ and usage of values Δ_f according to the data [33] are compared with experimentally estimated from a slope of the plots $N_{O_2}(t)$ in double log-log coordinates values d'^{e}_s in Figure 7 for mentioned above six tests of PAr and PAASO. As follows from the data of Figure 7, between values d'^{T}_s and d'^{e}_s the good correspondence is obtained, verified the conclusion about the necessity of usage of dimension d'_s instead of d_s at the description of high temperature oxidation reactions.

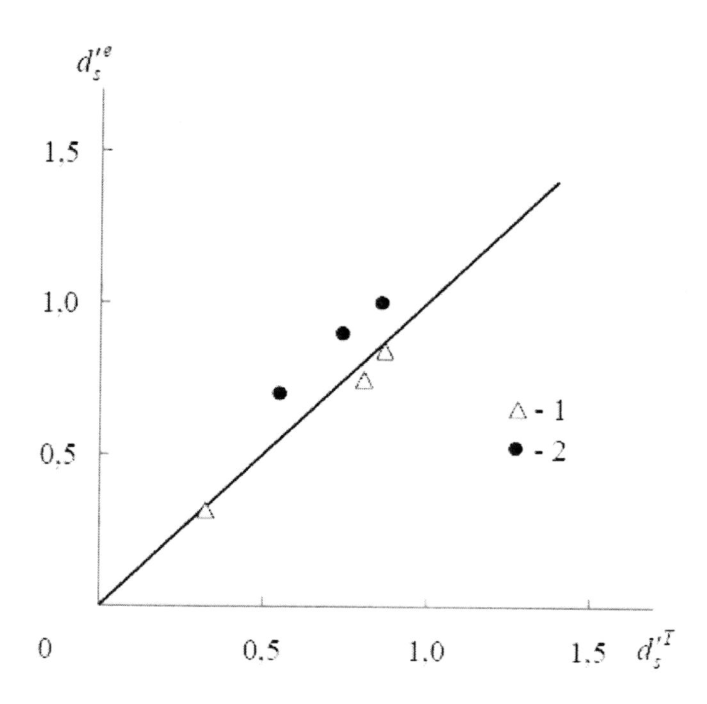

Figure 7. The comparison of calculated according to the equation (23) d'^{T}_s and determined experimentally d'^{e}_s effective spectral dimension values for PAr (1) and PAASO (2) [32].

Let's consider conditions of realization of autoaccelerated and autodecelerated kinetic curves according to the obtained above results. Boundary condition for the indicated transition is the criterion $x = 1$ in a relationship (12): at $x < 1$ the autodecelerated type of oxidation will be realized, and at $x > 1$ – an autoaccelerated mode. At $x = 1$ we shall obtain linear dependence $N_{O_2}(t)$, also occuring in practice [14]. For reactions such as (10) in Euclidean space the character of a curve $N_{O_2}(t)$ is defined by this space dimension: for $d = 1$ $x < 1$ (autodecelerated mode), for $d = 2$ $x = 2$ (linear dependence) and for $d = 3$ $x > 1$ (autoaccelerated regime). For reactions such as (11) in Euclidean space with any dimension $d \leq 3$ kinetic curves will represent autodecelerated mode. As the maximal values $d_s = 4/3$ and $d'_s = 1,20$ (the equation (23) at $\Delta_f = 3$), reactions in the fractal spaces will always be characterized by autodecelerated type of kinetic curves. The last conclusion was expected, as the curves of the mentioned type are characterized by decrease of reaction rate k with increase t (15).

As follows from relationship (19), the reactions such as (11) in an Euclidean space with d=3 are also characterized by kinetic curves of an autodecelerated type ($x = d/4 = 0,75$). In Figure 8 the comparison of two kinetic curves for reactions of a type (11), proceeding in Euclidean ($d = 3$) and fractal ($\Delta_f = 2,75$) spaces with the other equal conditions is shown. As it is possible to see, the reaction in an Euclidean space proceeds much faster, as it was expected [41].

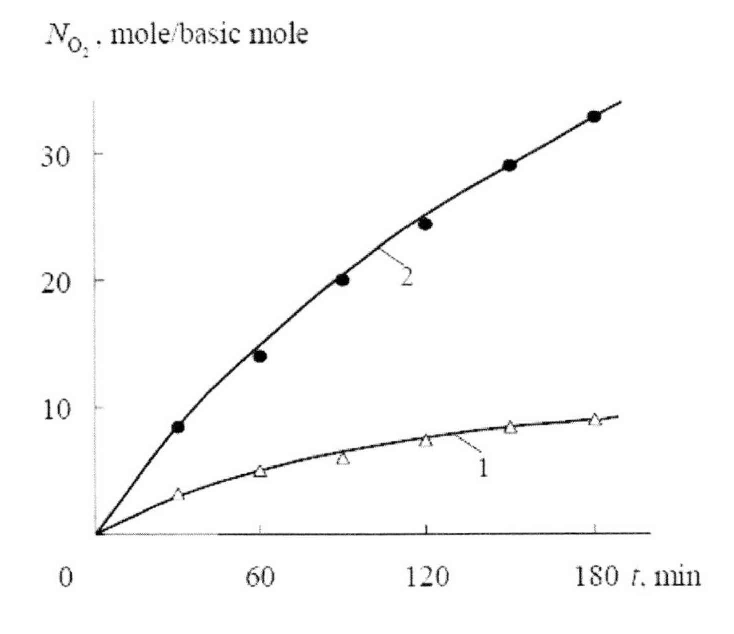

Figure 8. The comparison of kinetic curves $N_{O_2}(t)$ for PAASO-1 at $T = 623$ K for oxidation reactions proceeding in fractal ($\Delta_f = 2,75$) (1) and Euclidean ($d = 3$) (2) spaces [32].

The quoted above results allow the description of kinetic curves $N_{O_2}(t)$ with the help of the simple power or exponential laws. In Figure 5 the comparison of experimental and theoretical curves $N_{O_2}(t)$ is shown. The last curves are calculated according to following equations [42, 43]:

$$N_{O_2} = 0,67t^{d_s'/2}, \tag{24}$$

for PAASO-1 (autodecelerated type of curve) and

$$N_{O_2} = 0,0034t^{d/2} = 0,0034t^{1,5}, \tag{25}$$

for PAASO-2 (autoaccelerated type of curve). As it is possible to see, the equations (24) and (25) describe experimental kinetic curves $N_{O_2}(t)$ for both studied types correctly.

Therefore, the listed above results have shown, that the type of a kinetic curve of oxygen consumption in the process of a polymer melts oxidation is defined by the type of space, in which it proceeds, an in case of Euclidean space – its dimension *d*. The mentioned factors do not depend either on a chemical constitution of a polymer, or on testing temperature.

4. A Scaling Analysis of Polyurethanearylate Interphase Polycondensation with Varied Stirring Rate

Now is well known [44], that the intensive stirring of reactional medium in an interphase polycondensation process substantially raised its final characteristics (conversion degree Q and molecular weight M). This effect is explained by the fact that the increase of a splitting surface of nonmiscible phases because of the stirring accelerated diffusive processes and helped to polymer formation [44]. Within the framework of the scaling approach assumed, that the stirring decreases a non-homogeneity of medium which is due to large density fluctuations [45]. Let's consider this problem on an example of interphase polycondensation of polyurethanearylate (PUAr) [46].

In paper [45] we consider the reaction, in which particles P of a chemical substance diffuse in medium containing randomly located static nonsaturated traps (drowing macromolecular coils) T. At contact of a particle P with a trap T the particle disappears. Nonsaturation of a trap means that the reaction P + T → T can repeat inself infinite number of times. It is usually considered that the concentration of particles and traps is large or the reaction occurs at intensive stirring, the process can be considered as the classical reaction of the first order. In this case it is possible to consider that the law of the concentration decay of particles with time t will look like [45]:

$$c(t) \approx \exp(-At), \tag{26}$$

where A is constant.

However, if the concentration of the randomly located traps is small, then with the necessity there exist areas of space, practically free from traps. The particles getting into these areas, can reach the traps only during rather long time and, hence, the decay of their amount in the course of reaction will be slower. The formal analysis of this problem shows that the concentration of particles falls down under the law [45]:

$$c(t) \approx \exp\left(-Bt^{d/(d+2)}\right), \tag{27}$$

being dependent on the dimension of space d (B is constant).

If the traps can move, their mobility as though averages the influence of spatial heterogeneity, so the assumption resulting to (1) will be carried out better. In this case concentration of particles falls down under the combined law [45]:

$$c(t) \sim \exp\left(-At\right)\exp\left(-Bt\right)^{d/(d+2)}\Big). \tag{28}$$

However if to assume that the traps move very slowly as their mass more larger than particle mass and then the effect of their motion can be neglected [47].

The data of paper [46] have shown, that in an interval of a stirrer rotation speeds n= 4000÷6000 r.p.m. the value Q is reaching the asymptotic magnitude ~ 0,73. This means that the maximally possible effectivity of a fluctuation damping is reached corresponding to effective value d (d_{ef}) = 3,0. At minimal used stirrer rotation speed (1000 r.p.m) the value Q decreased approximately in 2,5 times. Therefore, it can be assumed an existence of density fluctuations and hence, an applicability of the equation (27). The estimation of the constants in the equations (26) and (27), i.e., A and B, from experimental data showed, that their absolute values are close and further their average magnitude, equal to 0,06 [48] will be used. To assume, that the exponent x in the equations (26) and (27) is varied linearly with n, then on the basis of listed above assumptions can be plotted, calibration plot shown in Fig. 9 $x(n)$. The value $c(t)$ can be expressed as $(1 - Q)$ and then for estimation Q the generalized equation [48] can be used:

$$1 - Q = \exp(-0{,}06\, t^{x}), \tag{29}$$

where the value x is determined from calibration plot of Figure 9 and value t = 180 min.

In Figure 10 the comparison of calculated according to the equation (29) and obtained experimentally [46] values Q is shown. As it is possible to see, between mentioned values Q a good correspondence is observed, that confirms correctness of the offered simple quantitative approach in case of stirring effect estimation [48].

At n = 0 (stirring absence) the value $x \approx 0{,}41$ is obtained, that is substantially lower than this exponent for d_{ef} = 3,0. Equaling the exponents in the equations (27) and (29), we can obtain for the case without stirring $d_{ef} \approx 1{,}39$. This means, that in indicated case interphase polycondensation proceeds in space intermediate between a line (d_{ef} = 1) and plane (d_{ef} = 2) in difference from volume space (d_{ef} = 3) in case of an intensive stirring. It can be said, that in case without stirring reaction proceeds on very perforated plane. Apparently, that in the first case (d_{ef} = 1,39) a number of contacts particle-trap will be substantially less than in the second one (d_{ef} = 3,0), that decreases sharply a reaction rate [48].

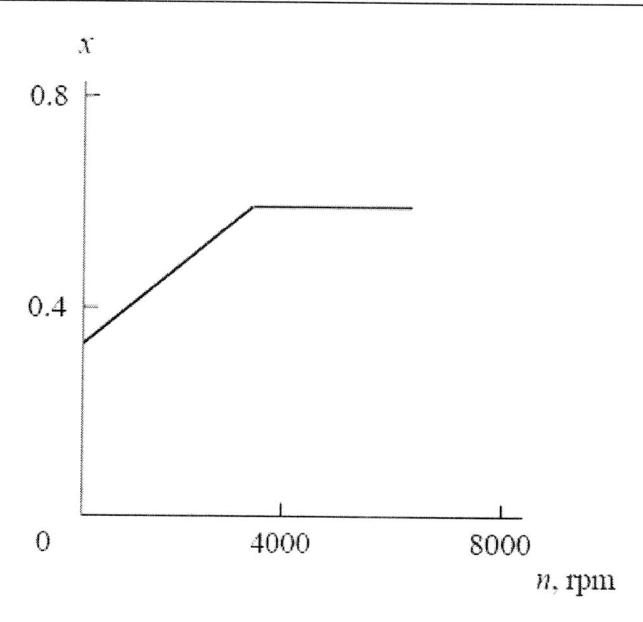

Figure 9. The calibration plot of the dependence of the exponent x in the equation (29) on stirring speed n. An explanations in the text are given [48].

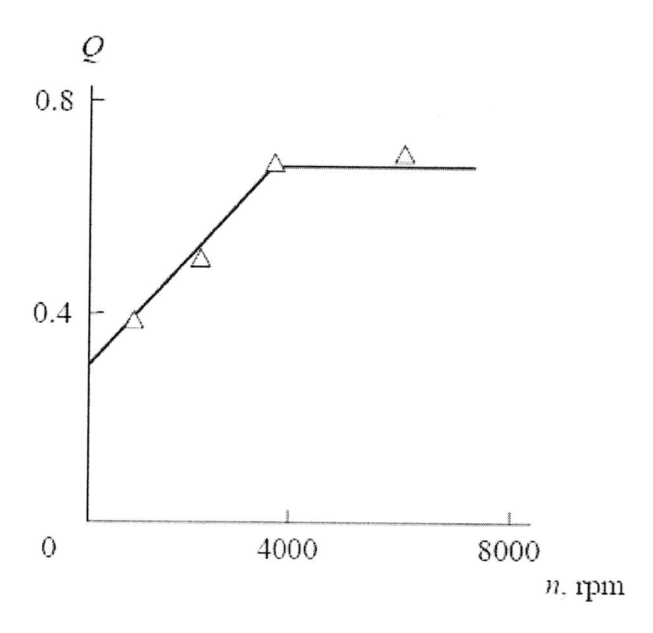

Figure 10. The theoretical (line) and experimental (points) dependences of a conversion degree Q on stirring rate n for PUAr [48].

In the framework of the irreversible aggregation model cluster-cluster the relationship between Q and M is expressed as follows [49]:

$$M \sim Q^{2/(3-D)}, \qquad (30)$$

where D is fractal dimension of polymer macromolecular coil in solution.

For polyarylates with enough high glass transition temperature obtained interphase polycondensation the value of exponent a in the Mark-Houwink-Sakurada equation is equal to ~ 0,75 [50]. Using this value a, D can be estimated according to the following equation [51]:

$$D = \frac{3}{1+a}, \tag{31}$$

that gives $D \approx 1{,}71$. Further, assuming in the first approximation $M \sim \eta_{red}$, where η_{red} is reduced viscosity of a polymer solution, we can estimate theoretical values η_{red} according to a relationship (31) and compare with the obtained experimentally [46]. Such comparison is shown in Figure 11, where a proportionality coefficient in a relationship (31) by method of fitting of theoretical and experimental dependences $\eta_{red}(n)$ is obtained. As follows from the data of Figure 11, offered treatment adequately described the dependence η_{red} (or M) on reactional medium stirring rate in case of interphase polycondensation of PUAr [48].

Hence, in polymers synthesis processes a substantional role plays not only fractal dimension D of macromolecular coil [52 – 54], but effective dimension of a space d_{ef}, in which the reaction proceeds. The less d_{ef} the less reacting molecules contacts and the lower main characteristics of synthesis process (Q and M). As mentioned above, the value d_{ef} can be determined equaling the exponents in the equations (28) and (30) [48]:

$$\frac{d_{ef}}{d_{ef}+2} = x. \tag{32}$$

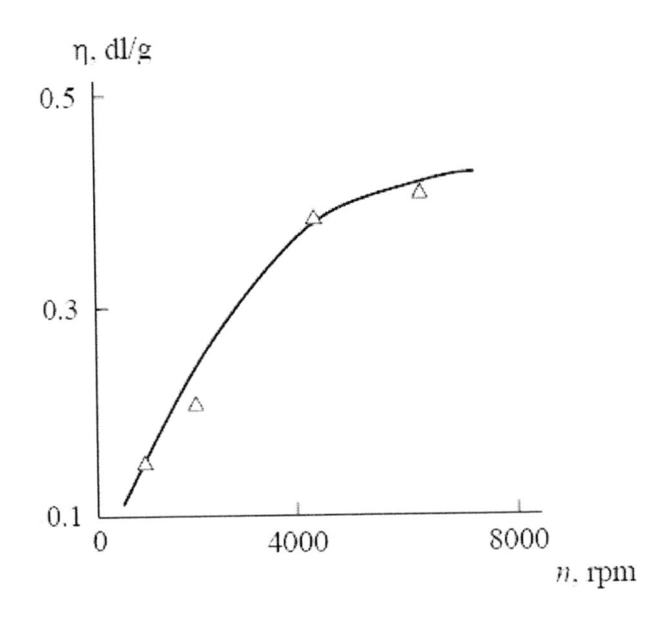

Figure 11. The theoretical (line) and experimental (points) dependences of a reduced viscosity η_{red} on stirring rate n for PUAr [48].

In Figure 12 the dependence d_{ef} on n for PUAr is shown. As follows from the data of this figure, at increase n from 0 up to 4000 r.p.m. the value d_{ef} increases from ~ 1,39 up to 3,0. Apparently, the values $d_{ef} > 3$ do not have physical sense [4]. The effective volume of medium V_{ef}, in which reaction proceeds, is a function of d_{ef} and can be written as follows [55]:

$$V_{ef} = L^{d_{ef}} \varepsilon^{3-d_{ef}}, \quad (33)$$

where L is the upper limit of fractal behaviour, ε is measurement scale.

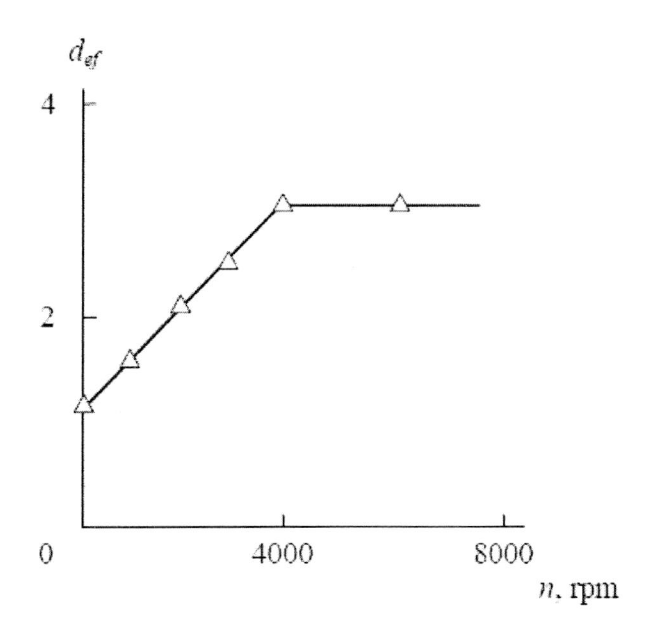

Figure 12. The dependence of effective dimension d_{ef} of space, in which one reaction proceeds, on stirring rate n for PUAr [48].

The estimation according to the equation (33) at arbitrary values L and ε shows large distinction of V_{ef} at d_{ef} = 1,39 and 3,0. So, at equal L = 10 and ε = 1 relative units the difference V_{ef} for cases without stirring and n = 4000 r.p.m. is equal to ~ 74 times. This estimation confirms the assumption about influence of stirring offered in paper [44].

Hence, a simple scaling approach allows to obtain not only qualitative, but quantitative description of the stirring effect on final characteristics of a polyurethanearylate interphase polycondensation. From physical point of view the interphase polycondensation rate is defined by a spatial fluctuations scale or effective (real) dimension of space, in which a relation proceeds [48].

5. The Fractal Analysis of Curing Processes of Haloidcontaining Epoxy Polymers

As the studies of curing kinetics of haloidcontaining epoxy polymers at different temperatures of curing T_{cur} showed there are two possible types of kinetic curves $Q(t)$ [56]. Examples of such different curves $Q(t)$ for two epoxy systems in Figure 13 (curves 1 and 2) are shown. As can be seen, for epoxy polymers on the basis of hexachloroetane (2DFP + HChE/DDM) a smooth decrease of the reaction rate is observed, whereas for epoxy polymers on the basis of hexachlorobenzene (HChB/DDM) the dependence $Q(t)$ is approximately linear up to $Q \approx 0,85$ and then curing reaction practically stops. Besides, as the data of paper [56] showed maximal degree of curing reaction conversion Q_{max} for epoxy polymers 2DFP + HchE/DDM is function of T_{cur} and varies in enough wide limits ($Q_{max} \approx 0,30 \div 0,85$), whereas for epoxy polymers HChB/DDM the value Q_{max} is practically independent from T_{cur} and reaches large absolute values (~ 0,95). Such obvious distinction of curves $Q(t)$ for mentioned epoxy polymers is considered in paper [57] in the framework of the fractal analysis and scaling approach. One of the possible methods of analytical description of curves $Q(t)$ is a following general fractal relationship [54]:

$$Q = K_1 c_0 \eta_0 t^{(3-D)/2}, \tag{34}$$

where K_1 is constant, c_0 is initial concentration of reagents, η_0 is initial viscosity of reactionary medium, D is fractal dimension of forming in curing process of cross-linked clusters (microgels).

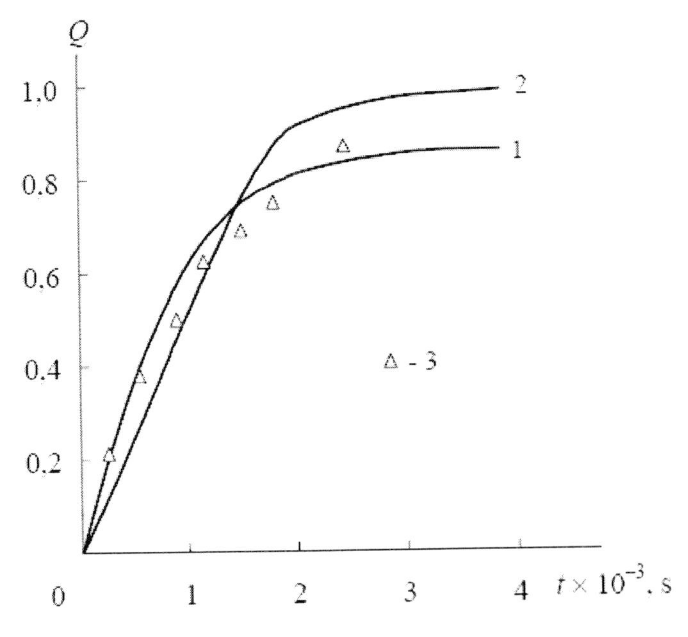

Figure 13. The experimental kinetic curves $Q(t)$ for epoxy polymers on the basis hexachloroethane (1) and hexachlorobenzene (2). 3 – the calculation according to the equation (34) [57].

The equation (34) described curves $Q(t)$ for linear polymers synthesis very well as in case of radical polymerization [58] and in case of polycondensation [54]. The value D can be determined from the slope of linear plot Q as function t in double log-log coordinates, as follows from the equation (34), and complex constant $K_1c_0\eta_0$ on – by fitting method. In Figure 13 such simulation for system 2DFP + HchE/DDM as points under conditions $K_1c_0\eta_0$ = const = $8{,}06 \times 10^{-6}$ and D = const = 1,78 is shown. As can be seen, good enough correspondence of experimental and theoretical curves is obtained up to t = 2400 s, where is the change of universality class of system due to gel formation and corresponding change of value D from 1,78 up to ~ 2,5 [59, 60]. To obtain an analogous description for system HChB/DDM is failed, as for it the value d is function of time t. In principle the curve 2 in Figure 13 can be described by usage variables d and η_0, but such approach is formal as the equation with two variables is described practically any smooth monotone curve.

Therefore, in paper [57] was made attempt to describe a curves $Q(t)$, shown in Figure 13, in the framework of scaling approach (the relationships (26) ÷ (28)). In Figure 14 the dependences ln(1 – Q) on t, corresponding to the relationship (26), for both mentioned systems. As follows from the shown plots, curing kinetics of system 2DFP + HChE/DDM enough well describes by linear correlation in coordinates of Figure 14, whereas the dependence [ln (1 – Q)](t) for system HChB/DDM linearize is failed. This means that the curing reaction for system 2DFP+HChE/DDM, described by the equation (34) at under mentioned above conditions is classical reaction of first order proceeding in reactionary medium with small density fluctuations [57].

The attempts to linearize the dependence (1 – Q) on t for system HChB/DDM with usage of the relationships (27) and (28) do not also lead to success. Therefore was made following assumption. The relationship (27) is described a reaction kinetics of low molecular weight substances at large density fluctuations in Euclidean space with dimension d (equal to 3 in considered case). If assume, that the formation of fractal clusters (microgels) with dimension d is defined course of curing reaction in fractal space with dimension d then dimension d in the relationship (27) should be replaced on D. The dependence ln(1 – Q) on $t^{D/(D+2)}$ corresponding to the relationship (27) with mentioned replacement for system HChB/DDM is shown in Fig. 15. Within this treatment the dependence $[\ln(1-Q)](t^{D/(D+2)})$ is linear and this circumstance indicated that the curing reaction of system HChB/DDM proceeds under conditions of large density fluctuations in a fractal space withe the dimension d [57].

Hence, the fractal reactions of polymerization can be divided, as a minimum, into two classes: reactions of fractal objects (homogeneous) whose kinetics is described similarly to the curve 1 shown in Figure 13, and reactions in fractal space (non homogeneous) whose kinetics is described similarly to the curve 2 in Figure 13. The reactions of the second class correspond to the formation of structures on fractal lattices [2, 3]. The basic distinction of the pointed classes of reactions is the dependence of their rate on fractal dimension d of products forming during the reaction (macromolecular coils, microgels). The first class of reactions is well described by the equation (34) (see Figure 13). The mentioned equation is obtained on the basis of theoretical conclusions of paper [61] where assume that the less d is the less compact is structure of fractal and the more sites are on it which are accessible to reaction. This means that the decreasing D leads to increase of reaction rate. In Figure 15 three modeling curves $Q(t)$, appropriate to the equation (34) with an identical constant for D = 1,5;

1,8 and 2,1 are shown. As it follows from the given curves, the increase D really sharply reduces the rate of reaction and decreases Q at the comparable values of t [57].

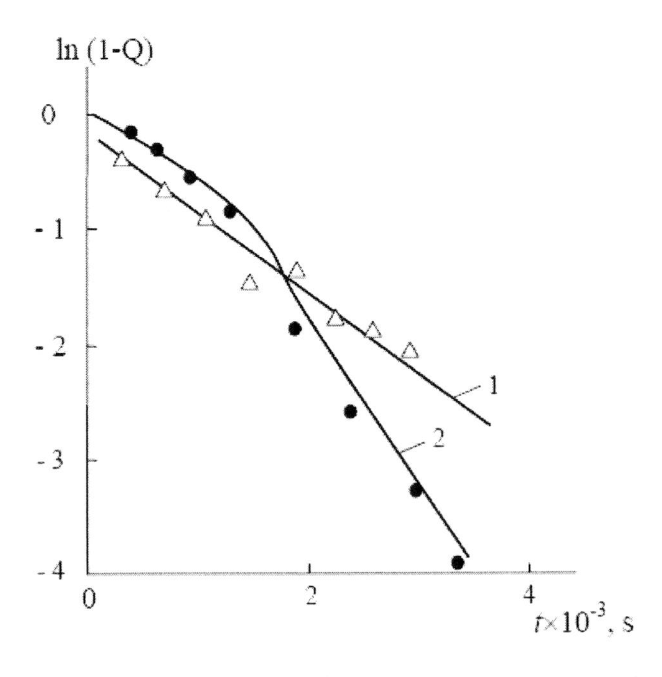

Figure 14. The dependences of $(1 - Q)$ on reaction duration t in double log-log coordinates corresponding to the equation (26) for systems 2DFP + HChE/DDM (1) and HChB/DDM (2) [57].

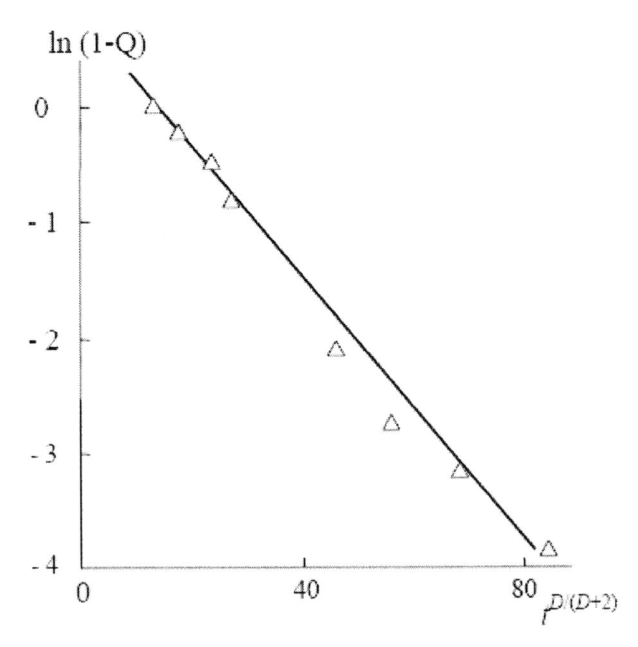

Figure 15. The dependence of $(1 - Q)$ on parameter $t^{D/(D+2)}$ in double log-log coordinates corresponding to the equation (27) for system HChB/DDM [57].

As to reactions in fractal spaces, there the situation is quite opposite. As it is known [10], if to consider a trajectory of diffusive movement of olygomer and curing agent molecules as a trajectory of random walks, the number of sites $\langle S \rangle$, visited by such walks, is determined according to the relationship (2) and defined by the value of spectral dimension d_s.As mentioned above, for Euclidean space $d_s = 3$ [9], for cured microgels $d_s = 1,33$ [9]. From a relationship (2) follows that the value $\langle S \rangle$ which can be treated as a number of contacts of reacting molecules, is proportional to $t^{1,5}$ in Euclidean and $t^{0,665}$ – in fractal spaces. At the identical t the greater number of the mentioned contacts in Euclidean space determines the faster curing reaction in comparison with a fractal space [57].

In this connection we shall note an interest detail. As it is shown in [62], for an ideal phantom network the following relationship is correct:

$$\frac{D}{D+2} = \frac{d_s}{2}. \tag{35}$$

It is easy see the obvious analogy between the parameters of the equation (27) (at replacement d on D) and a relationship (2).

In Figure 17 the curves $Q(t)$, calculated according to the equation (27) under the condition B = const for D = 1,5; 1,8 and 2,1 and also for d = 3 are given. It is easy to see that to in accordance with the above-stated treatment the rate of reaction increases at increase D and reaches the greatest value in Euclidean space at d = 3. It should be noted that in reactions of fractal objects according to a relationship (34) at $D = d = 3$ Q = const, and from view of a boundary condition $Q = 0$ at $t = 0$ is means that such reactions for three-dimensional Euclidean objects do not proceed at all [57].

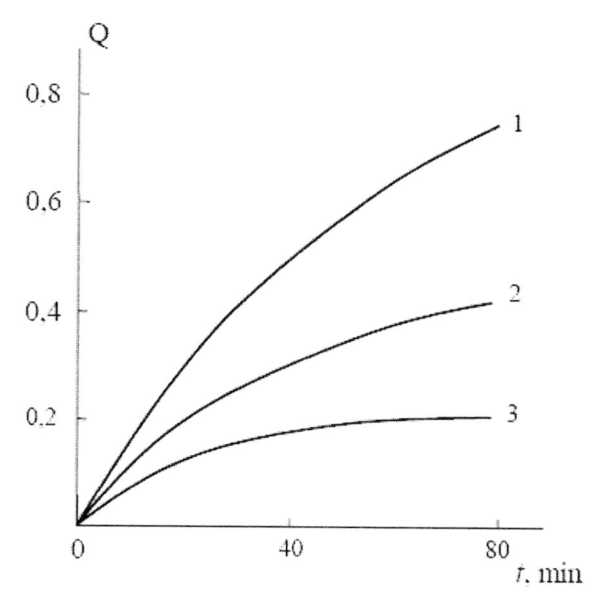

Figure 16. Modeling curves $Q(t)$ for reactions of the fractal objects calculated according to the equation (34) at D = 1,5 (1), 1,8 (2) and 2,1 (3) [57].

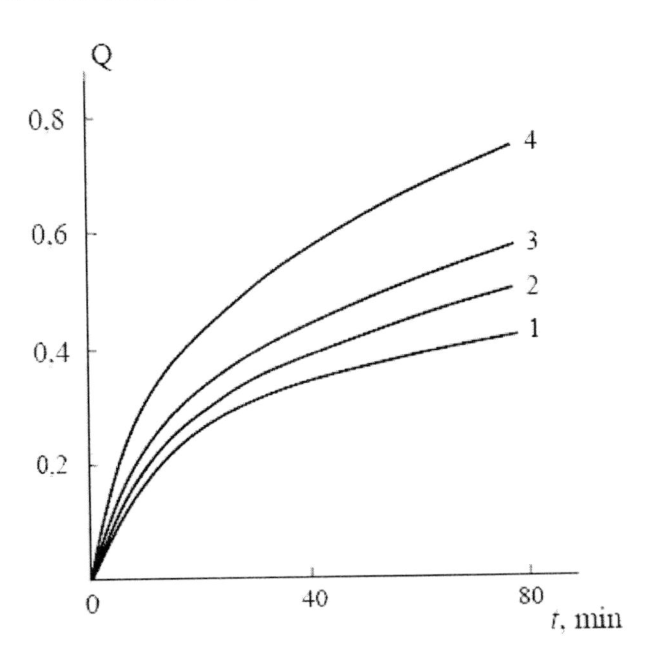

Figure 17. Modeling curves $Q(t)$ for reactions in the fractal space calculated according to relationship (27) at D = 1,5 (1), 1,8 (2), 2,1 (3) and d = 3 (4) [57].

6. The Theoretical Model of Volumetric Changes under Orientation of Polymers

The orientation of polymers by a uniaxial tension can be accompanied by essential volumetric changes. These changes are realized by various mechanisms defined by structure of polymeric material. So, at uniaxial drawing of semicrystalline high density polyethylene (HDPE) the change of volume up to 30 % due to the formation of cracks settled down perpendicularly to a drawing direction of polymerization-filled compositions (componors) on the basis of ultra-high-molecular polyethylene (HDMPE) the volume increase is due to the failure of interfacial boundaries a polymeric matrix-filler and can be reached ~ 10 % [64]. However, in case of a uniaxial tension of melt volumetric changes were not observed. It is assumed [63], that in this case at the orientation of melt the high mobility of macromolecules and the absence of the ordered elements of structure provide a tension of viscous medium without the formation of voids and the crystallization of such oriented melt results in the formation of system which doesn't contain voids and cracks. In paper [65] this behaviour was considered within the framework of the fractal analysis.

As Balankin [66] showed, the relative change of the excitation region volume in deformable body can be presented in the form:

$$\delta V_e = (1-2\nu_t)\frac{\sigma_t}{E_t} = (1-2\nu)\frac{\sigma}{E} \pm \delta V_s + \delta V_e, \tag{36}$$

were ν u ν_t are Poisson's ratios for initial and deformed polymer accordingly, σ and σ_e are fracture stresses for initial and deformed polymer accordingly, E and E_e are elastic modulus for initial and deformed polymer accordingly. The first member in the right part of the equation (36) is connected with elastic strans, the second one – with the relaxation of a stress by plastic deformation, the third one – with the formation of micro-, meso- and macrodefects. If the accumulation of the defects always results in the increase of volume, the change of volume connected with the plastic deformation has a sigh which opposite to an elastic component: "minus" at $\sigma > 0$ and "plus" at $\sigma < 0$ (compressive stress).

Dependence of Poisson's ratio ν_e of oriented polymer on the parameter

$$\Delta = \frac{\delta V_s \mp \delta V_e}{\delta V_e} \tag{37}$$

looks like [66]:

$$\nu_e = \frac{\nu + 0{,}5\Delta}{1+\Delta}. \tag{38}$$

The value ν for polymeric melt can be estimated as equal to 0,5. It is obvious that in this case $\nu_e = \nu$, as the value of Poisson's ratio cannot exceed 0,5 (the result of the principle of Le Shatelye-Brown) [66]. Then from the equation (38) it follows $\Delta = 0$ and the condition follows from the equations (36) and (37):

$$-\delta V_s + \delta V_e = 0. \tag{39}$$

Otherwise, the fractal analysis predicts the absence of the volumetric changes for any body with $\nu = 0{,}5$ (including the polymeric melt) owing to the compensation of volume changes caused by the formation of micro-, meso- and macrodefects, by volume changes caused by plastic deformation [65].

In case of uniaxial tension of solid semicrystalline polymer the situation changes. As the experimental estimations showed [67], the value of Poisson's ratio for initial polymeric materials (componors UHMPE-Al and UHMPE-bauxite) $\nu = 0{,}36$ and for extrudates of these materials with extrusion draw ratio $\lambda \geq 3$ one – $\nu_e \approx 0{,}43$. From the equation (38) it follows that $\Delta \approx 0{,}857$. It means the obligatory increase of componors volume expressed in formation of cracks on the interfaces polymeric matrix-filler [65]:

$$\delta V_e \Delta = -\delta V_s + \delta V_e. \tag{40}$$

As $\Delta \approx 0{,}857$ and $\delta V_e \approx 0{,}1$, from the equation (40) it follows that in this case plastic deformation cannot compensate the increase of volume caused by the formation of defects – cracks on the interfaces [65].

The fractal dimension d_f of the initial and d'_f of the oriented polymeric materials are connected by the following relationship [66]:

$$d'_f = \frac{d_f + 3\Delta}{1 + \Delta}. \tag{41}$$

As it was noted above, the value Δ is always more or equal to zero. It means, according to the equation (41), the increase of the fractal dimension of structure of a solid polymeric material at uniaxial tension that is experimentally confirmed [67, 68].

Thus, the listed above results have shown that the presence or absence of volumetric changes in process of uniaxial tension is caused by type of structure. If the structure is an Euclidean object (dimension $d = 3$; Poisson's ratio $\nu = 0{,}5$ [66]), the volumetric changes are absent; if it is a fractal object ($2 \le d_f < 2$; $0 \le \nu < 0{,}5$ [66]), the volumetric changes are obligatory [65].

7. A Formation of Particulate-Filled Composites Structure in Fractal Space

"The discturbation" of a polymeric matrix at incorporation of particulate filler within the framework of the fractal analysis is expressed as a fractal dimension d_f increase of its structure [69]. In papers [69] was also demonstrated, that a particles of particulate filler are formated in polymeric matrix backbone having fractal (in general case – multifractal) properties and characterized by fractal (Hausdorff) dimension D_b. Hence, the formation of a polymeric matrix structure in particulate-filled composite proceeds not in an Euclidean, but in a fractal space [71].

As it is known [72], the fractal dimension of an object is a function of the space dimension, in which it is formed. In computer simulation this situation is considered as the behaviour of fractals on fractal (instead of Euclidean) lattices [1 – 3]. The simple model of polymer macromolecular coil in the good solvents is the self-avoiding random, for which it is possible to written [2]:

$$R_g \sim N^{\nu_F}, \tag{42}$$

where R_g is a gyration radius of coil, N is a polymerization degree, ν_F is a Flory exponent.

On a fractal lattice with dimension D_{lat} the value ν_F can be estimated from the simple formula [2]:

$$\nu_F = \frac{3}{2 + D_{lat}}. \tag{43}$$

Assuming $D_{lat} = D_b$ (as it is mentioned above, we consider, that the polymeric matrix structure is formed not in an Euclidean three-dimensional but in a fractal space with dimension D_b [71]), it is possible to calculate values ν_F for a composites and then the fractal dimension of a macromolecular coil D according to the equation [37]:

$$D = \nu_F^{-1}. \tag{44}$$

The fractal dimension of polymer in the condensed state, i.e., value d_f, is possible to calculate as follows. The value D, which accouts for the excluded volume interactions, can be determined in this way [37]:

$$D = \frac{d_s(d+2)}{d_s+2}. \tag{45}$$

In present paper a composites polyhydroxiether-graphite (PHE–Gr) are studied. Two series of composites PHE–Gr are used: with the nonactivated (PHE–Gr-I) and the activated by a mixture of sulfuric and nitric acids in the ratio 1 : 1 on volume (PHE–Gr-II) filler. As PHE is a linear polymer, then for it it is necessary to accept $d_s = 1$ [9]. The fractal dimension of a polymer in the condensed state d_f is connected with d_s and d as follows [37]:

$$d_f = \frac{d_s(d+2)}{2}. \tag{46}$$

The combination of the equation (45) and (46) at $d_s = 1$ and any value d gives for the linear polymer [73]:

$$D_f = 1{,}5\, D. \tag{47}$$

In Figure 18 the comparison of the experimental d_f^e [69] and the calculated according to the equations (47) and (43) d_f^T values of the fractal dimension of a polymeric matrix structure for composites PHE–Gr-II is given in the form of dependence $d_f(\varphi_f)$, where φ_f is volumetric contents of a filler. As it is possible to see, the complete qualitative identity of the theoretical and experimental dependences $d_f(\varphi_f)$ is observed, however, the quantitative correspondence is much worse (the divergence between values d_f^e and d_f^T exceeds 15%). It is possible improve this correspondence empirically, having replaces factor 3 in the equation (43) on 2,5. In this case (see Figure 18) the excellent correspondence d_f^e and d'_f is obtained (the divergence is less than 1 %). The similar results are obtained for composites PHE–Gr-I, though for them the mentioned divergence is a little higher, but it does not exceed 4 %.

Nevertheless, the applicability of factor 2,5 in the equation (43) requires the theoretical confirmation.

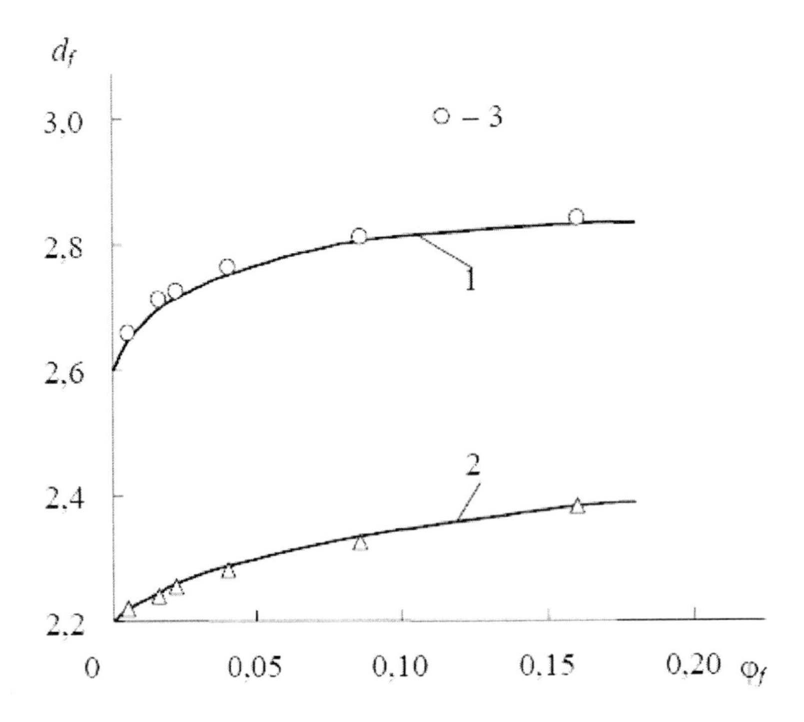

Figure 18. The dependences of the fractal dimension d_f of polymeric matrix structure on volumetric contents of a filler φ_f for composites PHE–Gr-II. 1 – the experimental data, 2,3 - calculation according to the equations (43) and (47) with usage in the equation (43) the coefficients 3 (2) and 2,5 (3) (see text) [71].

The most simple confirmation can be obtained for factor 2,5 in the equation (43) from the following boundary conditions. The values D_b and d_f in three-dimensional Euclidean space (d = 3) cannot be more by value d on the definition [74]. Substituting these limiting values D_b = 3 and d_f = 3 with account for a relationship (47) in the equation (43), we shall obtain factor 2,5 instead of 3 in a denominator of the mentioned equation.

For an estimation of value ν_F on fractal lattices the next formula can be also used, which is more exact [2]:

$$\nu_F = \frac{1}{D_{lat}} \cdot \frac{4d_e - d_s}{2 + 2d_e - d_s}, \tag{48}$$

where d_e is a chemical dimension or a dimension of "spreading", which characterizes the distance between two points of a fractal object not as geometrical distance (straight line connecting these points), but as "chemical distance", i.e., the shortest way on particles of an object between these two points [75].

To estimate value d_e we used the most simple variant based on the approximations of Flory and offered in paper [76]. For dimension $4/3 < d < 4$ it is possible to calculate the value of parameter Z:

$$Z = \frac{4+3D_b}{8}, \tag{49}$$

and then to determine d_e as follows [76]:

$$d_e = \frac{D_b}{Z}. \tag{50}$$

The spectral dimension d_s is determined according to the equation (20) and for calculation d_W can be used the approximation of Aharony-Stauffer [10]:

$$d_W = D_b + 1. \tag{51}$$

In Figure 19 the comparison of values d_f^e and d_f^T is given, where the theoretical magnitudes d_f are calculated by the considered above way with use of the equation (48). As it is possible to see, in this case not only a good qualitative, but good quantitative correspondence (maximal divergence of values d_f^e and d_f^T is less than 6 %) is also obtained. Let's note that the equation (48) is an appoximation and its check on Sierpinski gaskets has shown that the values ν_F, obtained according to this equation, diverged with the exact values of this parameter ~ 3,5 ÷ 7,5 % [2]. Such an error corresponds with the obtained in paper [71].

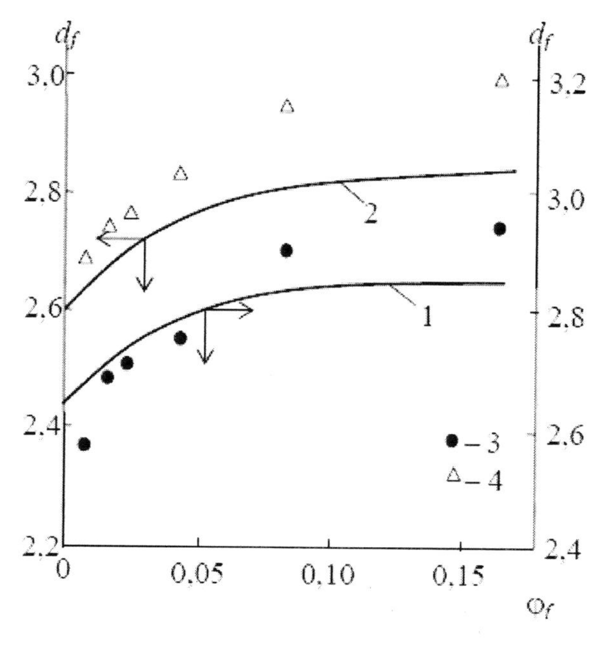

Figure 19. The dependences of the fractal dimension d_f of polymeric matrix structure on volumetric contents of a filler φ_f for composites PHE–Gr-I (1,3) and PHE–Gr-II (2,4). 1,2 – the experimental data, 3,4 – a calculation according to the equations (47) and (48) [71].

As it was shown in paper [2], the product $D_b\nu_F$ should be constant and independent of the distortions of a lattice (i.e., in present case – a backbone of filler particles), if its topology is fixed. In Figure 20 the dependence $D_b\nu_F(\varphi_f)$ is shown, where ν_F is estimated according to the equation (48), for the check of our calculations correctness. As follows from the data of Figure 20, the condition $D_b\nu_F$ = const is fulfilled with good precision, though some (less than 3 %) increase of this product at increase φ_f is observed. Taking into account the used approximations at calculations, it is necessary to admit such an error to be acceptable [71].

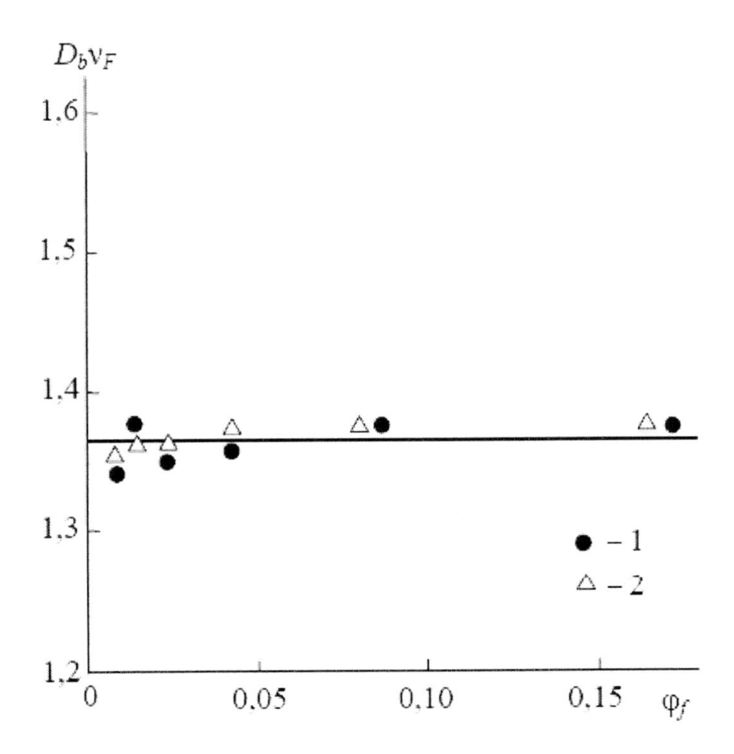

Figure 20. The dependence of a product $D_b\nu_F$ on volumetric contents of a filler φ_f for composites PHE–Gr-I (1) and PHE–Gr-II (2) [71].

As Meakin [77] assumed, the structural characteristics of interfacial layer are mainly determined by a dimension of space, in which its growth proceeds. For an interfacial layers was obtained the following relationship between a mean square thickness of interfacial layer l_{il} and particles number (statistical segments) in it N_i [77]:

$$l_{il} \sim N_i^{\varepsilon}, \tag{52}$$

where exponent ε = 1,7 for interfacial layers, which formation by a diffusive processes is controled, and ε = 1,0 one – for the conditions, where diffusive processes are insubstantial.

For composites PHE–Gr was obtained $\varepsilon \approx$ 1,10 [78] and for extrudates of polymerization filled compositions on the basis of UHMPE one – $\varepsilon \approx$ 1,13 [79]. This means, that the process of an interfacial layers formation in these materials practically without an influence of diffusive processes is proceed. Besides, this process is realized in the space with effective dimension less then 2 if allowed for that for d = 2 the exponent ε = 1,30 and for indicated above materials ε = 1,10 ÷ 1,13. In its turn, these structural changes are influenced definitely on a mechanical properties of these composites [80, 81].

Conclusions

The described in present paper examples are shown strong effect of a space type on reactions and structure formation for polymeric materials. A purposeful change of a space type will be allow sharply vary reaction rate depending on that is this reaction wishing (synthesis) or not (degradation). The offered technique allows also simulate and predict a structure of composites polymeric matrix depending on conditions of a initial filler aggregation.

References

[1] Meakin P. *Phys. Rev.* B, 1984, v. 29, No. 8, p. 4327-4330.
[2] Vannimenus *J. Physica* D, 1989, v. 38, No. 2, p. 351-355.
[3] Aharony A., Harris A.B. *J. Stat. Phys*., 1989, v. 54, No. 3/4, p.1091-1097.
[4] Kozlov G.V., Novikov V.U. Synergetics and Fractal Analysis of Cross-Linked Polymers. (in Russian) Moscow, *Klassika,* 1998, 112 p.
[5] Rammal R., Toulouse G. *J. Phys. Lett*. (Paris), 1983, v. 44, No.1, p. L13-L22.
[6] Afaunov V.V., Kozlov G.V., Mashukov N.I., Zaikov G.E. *Zhurnal Prikladnoyi Khimii*, 2000, v. 73, No. 1, p. 136-140.
[7] Mashukov N.I., Kozlov G.V., Afaunof V.V., Zaikov G.E. Intern. Conf. MoDeSt-2000, Sept. 3-7, 2000, University of Palermo, Palermo, Italy, 2000, P-1/w/20.
[8] Meakin P., Stanley H.E. *Phys. Rev. Lett*., 1983, v. 51, No. 16, p. 1457-1460.
[9] Alexander S., Orbach R. *J.Phys. Lett*. (Paris), 1982, v. 43, No. 17, p. L625-L631.
[10] Sahimi M., McKarnin M., Nordahl T., Tirrell M. *Phys. Rev.* A, 1985, v. 32, No. 1. P. 590-595.
[11] Mashukov N.I., Gladyshev G.P., Kozlov G.V. *Vysokomolekulyarnye Soedineniya*, A, 1991, v. 33, No. 12, p. 2538-2546.
[12] Mashukov N.I. A Stabilization and Modification of High Density Polyethylene by Oxygen Acceptors. Diss… dokt. *Khim. Nauk*, Moscow, MkhTI, 1991, 422 p.
[13] Mashukov N.I., Vasnetzova O.A., Kozlov G.V., Kesheva A.B. *Lakokrasochnye materialy i ikh primenenie,* 1990, No. 5, p. 38-41.
[14] Kovarskaya B.M., Blumenfeld A.P., Levantovskaya I.I. Thermal Stability of Heterochain Polymers (in Russian), Moscow, *Khimiya*, 1977, 264 p.
[15] Kopelman R. In book: *Fractals in Physics*. Ed. Pietronero L., Tosatti E. Amsterdam, Oxford, New York, Tokyo, North-Holland, 186, p. 524-527.
[16] Kozlov G.V., Mashukov N.I., Zaikov G.E., Mikitaev A.K., Borukaev T.A. In book: *Fractals and Local Order in Polymeric Materials*. Ed. Kozlov G., Zaikov G. New York, Nova Science Publishers, Inc., 2001, p. 11-19.
[17] Borukaev T.A., Kozlov G.V., Mashukov N.I., Mikitaev A.K. *Plasticheskie massy*, 2002, No. 9, p. 25-26.

[18] Klymko P.W., Kopelman R. *J. Phys. Chem.*, 1983, v. 87, No. 23, p. 4565-4567.

[19] Kozlov G.V., Shogenov V.N., Afaunov V.V., Zaikov G.E. *Oxidation Commun.*, 2003, v. 26, No. 1, p. 121-126.

[20] Kalinchev E.L., Sakovtseva M.B. Properties and processing of thermoplastics (in Russian). Leningrad, *Khimiya*, 1983, 288 p.

[21] Kozlov G.V., Novikov V.U. *Uspechi Fizicheskich Nauk*, 2001, v, 171, No. 7, p. 717-764.

[22] Shogenov V.N., Gazaev M.A., Kardanov M.Sh. In book: *Polycondensational Processes and polymers*. Ed. Korshak V.V., Nal'chik, KBSU, 1987, p. 8-12.

[23] Kozlov G.V., Dolbin I.V., Zaikov G.E. In book: *Aging of Polymers, Polymer Blends and Polymer Composites*. V. 2. Ed. Zaikov G., Buchachenko A., Ivanov V. New York, Nova Science Publishers, Inc., 2002, p. 145-150.

[24] Emanuel N.M. *Vysokomolekulyarnye Soedineniya*, A, 1985, v. 27, No. 7, p. 1347-1365.

[25] Kozlov G.V., Shustov G.B., Zaikov G.E. In book: *New Perspectives in Chemistry and Biochemistry*. Ed. Zaikov G. New York, Nova Science Publishers, Inc., 2002, p. 11-17.

[26] Kozlov G.V., Shustov G.B., Zaikov G.E. In book: *Chemistry and Biochemistry on the Leading Edge*. Ed. Zaikov G. New York, Nova Science Publishers, Inc., 2002, p. 21-29.

[27] Grassberger P., Procaccia I. *J. Chem. Phys.*, 1982, v. 77. No. 12, p. 6281-6284.

[28] Kang K., Redner S. *Phys. Rev. Lett.*, 1984, v. 52, No. 12, p. 955-958.

[29] Meakin P., Stanley H.E. *J. Phys.* A, 1984, v. 17, No. 1, p. L173-L177.

[30] Redner S., Kang K. *J. Phys.* A, 1984, v. 17, No. 3, p. L451-L455.

[31] Blumen A., Klafter J., Zumofen G. In book: *Fractals in Physics*. Ed. Pietronero L., Tosatti E. Amsterdam-Oxford-New York-Tokyo, North-Holland, 1986, p. 561-574.

[32] Kozlov G.V., Novikov V.U., Zaikov G.E. In book: *Reactions in Condensed Phase. Kinetics and Thermodynamic*. Ed. Zaikov G. New York, Nova Science Publishers, Inc., 2003 (in press).

[33] Kozlov G.V., Shustov G.B., Zaikov G.E. In book: *Aging of Polymers, Polymer Blends and Polymer Composites*. V. 2. Ed. Zaikov G., Buchachenko A., Ivanov V. New York, Nova Science Publishers, Inc., 2002, p. 151-160.

[34] Kozlov G.V., Dolbin I.V., Zaikov G.E. In book: *Reactions in Condensed Phase. Kinetics and Thermodynamics*. Ed. Zaikov G. New York, Nova Science Publishers, Inc., 2003 (in press).

[35] Bershtein V.A., Egorov V.M. Differential Scaling Calorimetry in Physico-Chemistry of Polymers (in Russian). Leningrad, *Khimiya*, 1990, 256 p.

[36] Lobanov A.M., Frenkel S.Ya. *Vysokomolekulyarnye Soedineniya*, A, 1980, v. 22, No. 5, p. 1045-1057.

[37] Vilgis T.A. *Physica* A, 1988, v. 153. No. 2. P. 341-354.

[38] Shlyapnikov Yu.A., Kiryushkin S.G., Mar'in A.P. *Antioxidative Stabilization of Polymers*. London, Taylor and Francis, 1996, 256 p.

[39] Dolbin I.V., Kozlov G.V., Zaikov G.E. In book: *Abstracts of International Symposium "Fractals and Applied Synergetics,* FiPS-2001". Moscow, MSOU, 2002, p. 41-42.

[40] Kozlov G.V., Batyrova H.M., Zaikov G.E. In book: *New Perspectives in Chemistry and Biochemistry*. Ed. Zaikov G. New York, Nova Science Publishers, Inc., 2002, p. 49-55.

[41] Shogenov V.N., Kozlov G.V. *The Fractal Clusters in Physico-Chemistry of Polymers*. Nal'chik, Polygraphservice and T, 2002, 268 p.

[42] Kozlov G.V., Zaikov G.E., Lipatov Yu.S. *Doklady NAN Ukraine*, 2002, No. 8, p. 130-135.
[43] Kozlov G.V., Zaikov G.E., Lipatov Yu.S. *Russian Polymer News*, 2003, v. 8, No. 1, p. 59-63.
[44] Korshak V.V., Vinogradova S.V. *Nonequilibrium Polycondensation* (in Russian). Moscow, Nauka, 1972, 695 p.
[45] Djordjevic Z.B. In book: *Fractals in Physics*. Ed. Pietronero L., Tosatti E. Amsterdam-Oxford-New York-Tokyo, North-Holland, 1986, p. 581-585.
[46] Mikitaev A.K., Korshak V.V., Afaunova Z.I. *Vusokomolekylyarnye Soedineniya* A, 1972, v. 14, No. 10, p. 2111-2114.
[47] Koeb M. *J. Phys.* A, 1986, v. 19, No. 5, p. L263-L268.
[48] Afaunova Z.I., Kozlov G.V., Vestnik KBSU, *Fizicheskie Nauki.*, 2000, No. 5, p. 48-53.
[49] Kozlov G.V., Temiraev K.B., Ovcharenko E.N., Lipatov Yu.S. *Doklady NAN Ukraine*, 1999, No. 12, p. 136-140.
[50] Askadckii A.A. Physico-Chemistry of Polyarylates (in Russian). Moscow. *Khimiya,* 1968, 214 p.
[51] Karmanov A.P., Monakov Yu.B. *Vysokomolekulyarnye Soedineniya* B, 1995, v. 37, No. 2, p. 328-331.
[52] Shiyan A.A. *Vysokomoleculyarnye Soedineniya* B, 1995, v. 37, No. 9, p. 1578-1580.
[53] Kozlov G.V., Temiraev K.B., Kaloev N.I. *Doklady Russian AN*, 1998, v. 362, No. 4, p. 489-492.
[54] Novikov V.U., Kozlov G.V. *Uspehi Khimii*, 2000, v. 69, No. 6, p. 572-599.
[55] Van Damme H., Levitz P., Bergaya F., Alkover J.F., Gatineau L., Fripiat J.J. *J. Chem. Phys.*, 1986, v. 85, No. 1, p. 616-625.
[56] Kozlov G.V., Bejev A.A., Zaikov G.E. *Oxidation Commun.*, 2002, v. 25, No. 4, p. 529-534.
[57] Kozlov G.V., Bejev A.A., Lipatov Yu.S. In book: *Perspectives on Chemical and Biochemical Physics*. Ed. Zaikov G. New York, Nova Science Publishers, Inc., 2002, p. 231-253.
[58] Kozlov G.V., Ozden S., Malkanduev Yu.A., Zaikov G.E. *Russian Polymer News*, 2002, v. 7, No. 3, p. 38-44.
[59] Botet R., Jullien R., Kolb M. *Phys. Rev.* A, 1984, v. 30, No. 4, p. 2150-2152.
[60] Kobayashi M., Yoshioka T., Imai M., Itoh Y. *Macromolecules*, 1995, v. 28, No. 22, p. 7376-7385.
[61] Pfeifer P., Avnir D., Farin D. *J. Stat. Phys.*, 1984, v. 36, No. 5/6, p. 699-716.
[62] Hess W., Vilgis T.A., Winter H.H. *Macromolecules*, 1988, v. 21, No. 8, p. 2536-2542.
[63] El'yashevich G.K., Karpov E.A., Lavrent'ev V.K., Poddubnyi V.I., Genina M.A., Zabashta Yu.F. *Vysokomolekulyarnye Soedineniya* A, 1993, v. 35, No. 6, p. 681-685.
[64] Kozlov G.V., Beloshenko V.A., Slobodina V.G. *Plasticheskie massy*, 1996, No. 3, p. 14-16.
[65] Aloev V.Z., Kozlov G.V., Dolbin I.V., Zaikov G.E. In book: *Perspectives on Chemical and Biochemical Physics.* Ed. Zaikov G. New York, Nova Science Publishers, Inc., 2002, p. 175-177.
[66] Balankin A.S. *Synergetics of Deformable Body* (in Russian). Moscow, MO SSSR, 1991, 404 p.

[67] Kozlov G.V., Beloshenko V.A., Varyukhin V.N., Gazaev M.A. *Prikladnaya Mekhanika i Technicheskaya Fizika,* 1998, v. 39, No. 1, p. 160-163.
[68] Kozlov G.V., Beloshenko V.A., Varyukhin V.N., Novikov V.U. *Zhurnal Fizicheskikh Issledovanii,* 1997, v. 1, No. 2, p. 204-207.
[69] Kozlov G.V., Mikitaev A.K. *Mechanika Kompozitsionnykh Materialov i Konstruktsii,* 1996, v. 2, No. 3-4, p. 144-157.
[70] Novikov V.U., Kozlov G.V. *Mechanika Kompozitnykh Materialov,* 1999, v. 35, No. 3, p. 269-290.
[71] Kozlov G.V., Yanovskii Yu.G., Lipatov Yu.S. *Mekhanika Kompozitsionnykh Materialov i Konstruktsii,* 2002, v. 8, No. 4. P. 467-474.
[72] Tokuyama M., Kawasaki K. *Phys. Lett.,* 1984, v. 100A, No. 7, p. 337-340.
[73] Kozlov G.V., Temiraev K.B., Shustov G.B., Mashukov N.I. *J. Appl. Polymer Sci.,* 2002, v. 85, No. 6, p. 1137-1140.
[74] *Feder E. Fractals.* New York, Plenum Press, 1989, 256 p.
[75] Vannimenus J., Nadal J.P., Martin H. *J. Phys.* A, 1984, v. 17, No. 6, p. L351-L356.
[76] Lhuillier D. *J. Phys. France,* 1988, v. 49, No. 5, p. 705-710.
[77] Meakin P. *Phys. Rev.* A, 1983, v. 27, No. 5, p. 2616-2623.
[78] Burya A.I., Shogenov V.N., Kozlov G.V., Kholodilov O.V. *Materialy, Technologii, Instrumenty,* 1999, v. 4, No. 2, p. 39-41.
[79] Aoev V.Z., Kozlov G.V. *Fizika i Technika Vysokikh Davlenii,* 2001, v. 11, No. 1, p. 40-42.
[80] Beloshenko V.A., Kozlov G.V., Varyukhin V.N., Slobodina V.G. *Acta Polymerica,* 1997, v. 48, No. 5-6, p. 181-192.
[81] Mashukov N.I., Novikov V.U., Kozlov G.V., Bur'yan O.Yu. *Materialovedenie,* 2000, No. 3, p. 35-37.

In: Reactions and Properties of Monomers and Polymers ISBN: 1-60021-415-0
Editors: A. D'Amore and G. Zaikov, pp. 199-204

Chapter 10

METHODOLOGY OF DEVELOPMENT AND ESTIMATION OF QUALITY OF HEAT-RESISTANT ANTIADHESIVE COATINGS ON THE BASIS OF FUSIBLE FLUOROPLASTICS FOR FOOD INDUSTRY

V. V. Anan'ev, M. I. Gubanova* and G. V.Semenov
Moscow State University of Applied Biotechnology (MSUAB)

ABSTRACT

Development of antiadhesive coatings for working surfaces of the food equipment in high-temperature technology is extremely important for the modern food industry. By development of the coatings as a polymeric basis a powder fluoroplastic modified by the complex of additives is used. The studies are performed on estimation of modifying component distribution in a polimer matrix.

At present time the complexity of the problem of the development of antiadhesive coatings for alimentary purposes is conditioned by restricted selection of polymer materials suitable for food-contacting covers which possess high thermostability, and also low surface energy responsible for a level of adhesive interplay with alimentary media. Therefore a problem of widening of the range of thermo- and chemoresistant polymer materials by usage of number of unsoluble biologically inert polymers as film-forming components is actual.

Widely known for today bakers' coatings developed on the basis of suspension fluoroplastics[1-7] and organosilicon compounds are intended mainly for exploitation in mild conditions: on aggregates of small and medium powers of a periodic type, in regimens excluding hot idle times and thermal shocks and not intended for exploitation conditions on native-country high-performance aggregates of continuous operating, including periods of inexact and spasmodic load. Besides that, the coatings known up-today technologically do not

* Corresponding author: Talalihina street, 33 109316 Moscow Russia E-mail: m_guban@rambler.ru

allow to create solid uniform-thickness antiburnfasting coverings neither for the grain moulds, nor for perforated sheets, corrugated confectionery moulds, and profiles of complex configuration. Disadvantages of known domestic and foreign coatings on the basis of suspension compositions are: labourious input and long duration of the process of manufacturing, and also usage in this process of ecologically hazard materials (toluene, xylene etc.). The coating on the basis of solution systems do not provide an equal bed depth on the all surface of the equipment, that in turn generates temperature stresses and thermal deformations of covering, appearance of unequal conditions of heating on separate segments of the equipment with its subsequent buckling or burn-out and other undesirable consequences. To the greatest degree the elimination of the indicated negative factors is possible only usining powder antiburnfasting coatings.

Thermal and chemical resistance of the fluoroplastics, their nonwettability with water, inertness to fats, oils, organic solvents, and also positive sanitary-hygienic characteristics allowed to recommend them for use in baking industry [8,9]. From powder fluoroorganic polymers suitable for a contact to food, Fluoroplast-4MB has the most valuable protective attributes. However, because of large melt viscosity at high temperatures [9], absence of a viscous-flow condition and predilection of the powder to balling up, manufacturing of coatings from Fluoroplast-4MB is connected with large technical difficulties.

In this connection the problem is actual of development of competitive compositions for antiadhesive coating on the basis of powder fluoroplastics satisfying to such requirements as:

- ecological and biological safety;
- maintenance of stable high separating effect, i.e. minimum adhesion to a product and maximum adhesion to metallic surfaces of the equipment or devices;
- thermostability, stability to thermal shocks, preservation of functionability at alternating temperatures;
- non-toxicity, chemical and physiological inertness, absence of migration of coating components into food and environment;
- manufacturability, capability of deposition by a homogeneous layer on the surface of complex configuration, reproducibility of parameters;
- high level of physical-mechanical properties providing long-lived exploitation of coatings in a given temperature range: strength, elasticity, abrading resistance, durability etc.;
- technical, social and economic efficiency of use.

In MSUAB during a number of years the complex researches were carried out concerning with the technology of powder polymer compositions and multifunctional protective coatings on this basis. Recently, coating heat stabilisers and two generations of antiburnfasting coatings predominantly used for deposition on the bakers' mould internal surfaces and other technological equipment of baking industry were designed, approved and introduced in the food-processing industry. A distinctive feature of exploitation of the baking industry is the very rigid requirements to thermal, mechanical and adhesive stability of antiburnfasting coatings. Within the framework of further development of studies on creation of a new generation of separating coatings with improved functional properties we have carried out the researches, as a result of which new modifying complexes for powder fluoroplastic compositions were developed.

As contrasted to precursor compositions on the basis of powder fluoroplastics, modifying agents are excluded from their content, because their presence results in cover cracking and loss of antiburnfasting properties at hardening of operation conditions. Toxic components (for example, triphenylphoshine) also are excluded. The compositions are modified with the components providing increase in adhesion strength of a powder coating to a surface of metal shapes with preservation these properties even after numerous cyclic heating of the shapes to high temperatures.

In collaboration with the Institute of Physical Chemistry of Russian Academy of Sciences the structure of new coatings was studied by the method of electronic microscopy. The study was executed with the scanning microscope "Philips" equipped with an electron probe analyzer "Kevex". Use of the microanalyzer allowed to evaluate the distribution of modifying components in the fluoroplastic compositions. Samples of films formed on a neutral (glass) support, and also samples of coatings on a metal support (silumin) were studied. The surface of coatings, and the structure of cross-sectional shears of films and metal-supported coatings were also investigated. Cross-sectional shears were prepared with use of a ultramicrotome. The microscope permission at this study was 60 Angstroems.

In the figure 1 a cross-sectional shear of the silumin support (left-hand light part) with the coating and a characteristic distribution of basic elements (fluorine, aluminum and titanium) on the scanning pathway of the cross-sectional shear are shown.

Basing on the results of the electronic-microscopic analysis, it have been possible to conclude that the defects on the border aluminum - fluoroplastic are not revealed.

The intensity (I) of a secondary X-irradiation characteristic for the given element is shown on an ordinate axis of the graphs in relative units, and a position coordinate (X) of a scanned sample site is shown on an abscissa axis, assuming a border between the support and the coating as zero point.

The appearance of irradiation from fluorine and titanium is observed in the same coordinates. The relation of a mean emission power to the scanning point coordinate allows to conclude on an uniform distribution of these elements on the sample depth and, therefore, on the component distribution in the bulk of the sample.

The analysis of microphotographs obtained from the surface of films and coatings, and also from their cross-sectional shears (the analysis was performed on the following element: nitrogen, boron, titanium, chromium, oxygen), convinces that the fluoroplastic coating with the offered modifying components has rather homogeneous structure and is characterized by an uniform distribution of the additives. Such structure, probably, occurs in the results of not only component composition, but also of the regimen of coating formation, and provides the stability of properties during cyclic exposures to high temperatures.

Together with the Research Institute of Plastics we studied free films of the initial fluoroplastic and of two modified fluoroplastic compositions by the method of IR-spectroscopy. Spectra were recorded on the IR-Fourrier spectrophotometer "Spectrum One" of the corporation Perkin-Elmer by a method «on a lumen», and also with use of an attachment "Universal ATR" by a method of disturbed total internal reflection (DTIR). In the latter case the spectra were identical for all three tested samples: there were absorption bands in the wave number regions of 1200, 1150, 980 and 750 cm^{-1} (figure 2).

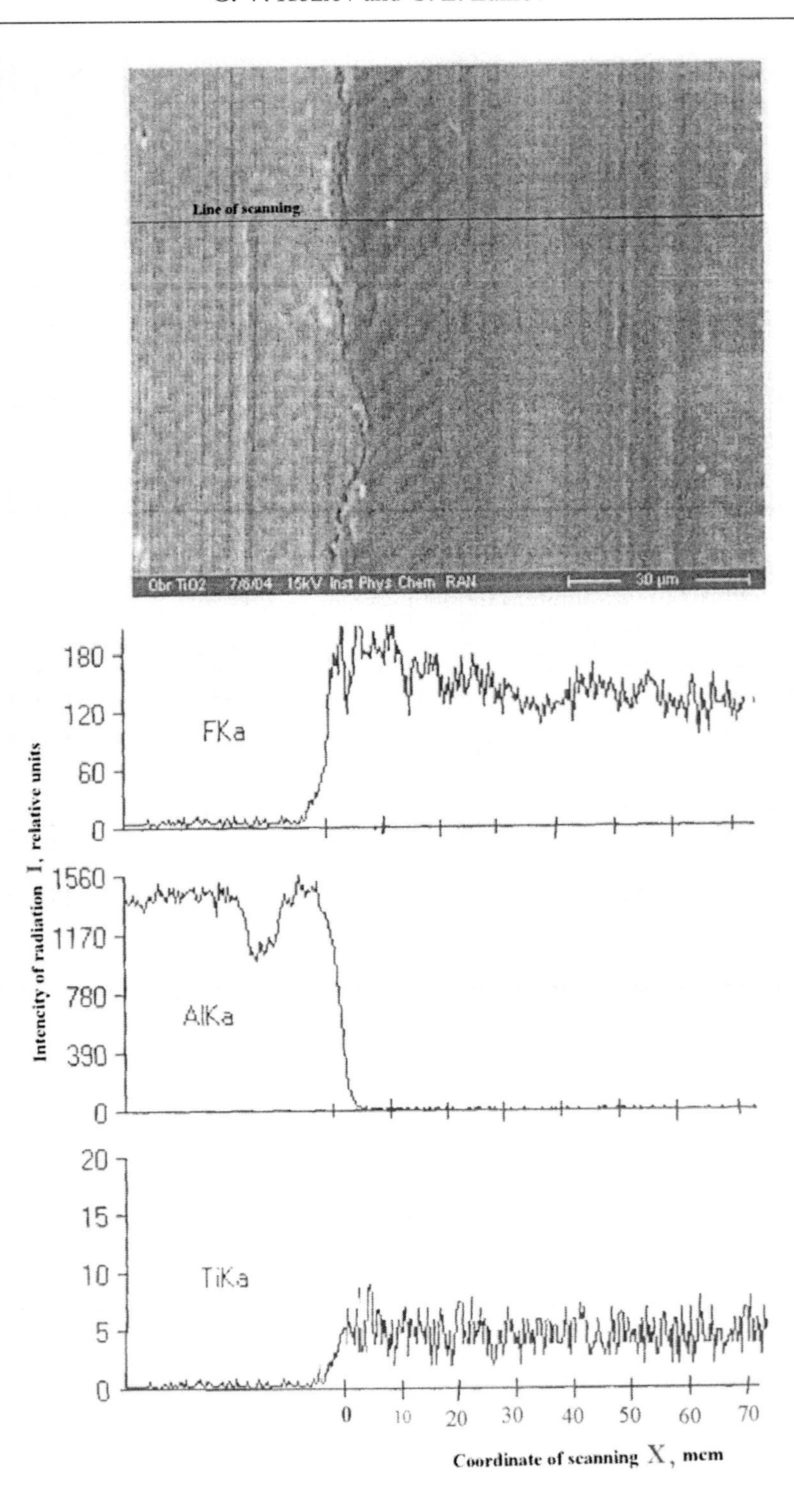

Figure 1.

The observed effect could be explained to that the upper layer of the coating (depth less than 1 micron) is mpoverished with modifying components, and their main part is distributed in the bulk of the sample and on the border with the support. At such distribution of modifying components in the coating the danger of their migration at the contact with food is considerably reduced, the physiological inertness of fluoroplastic is saved, and simultaneously the minimal level of adhesive interaction between the coating and the product is provided.

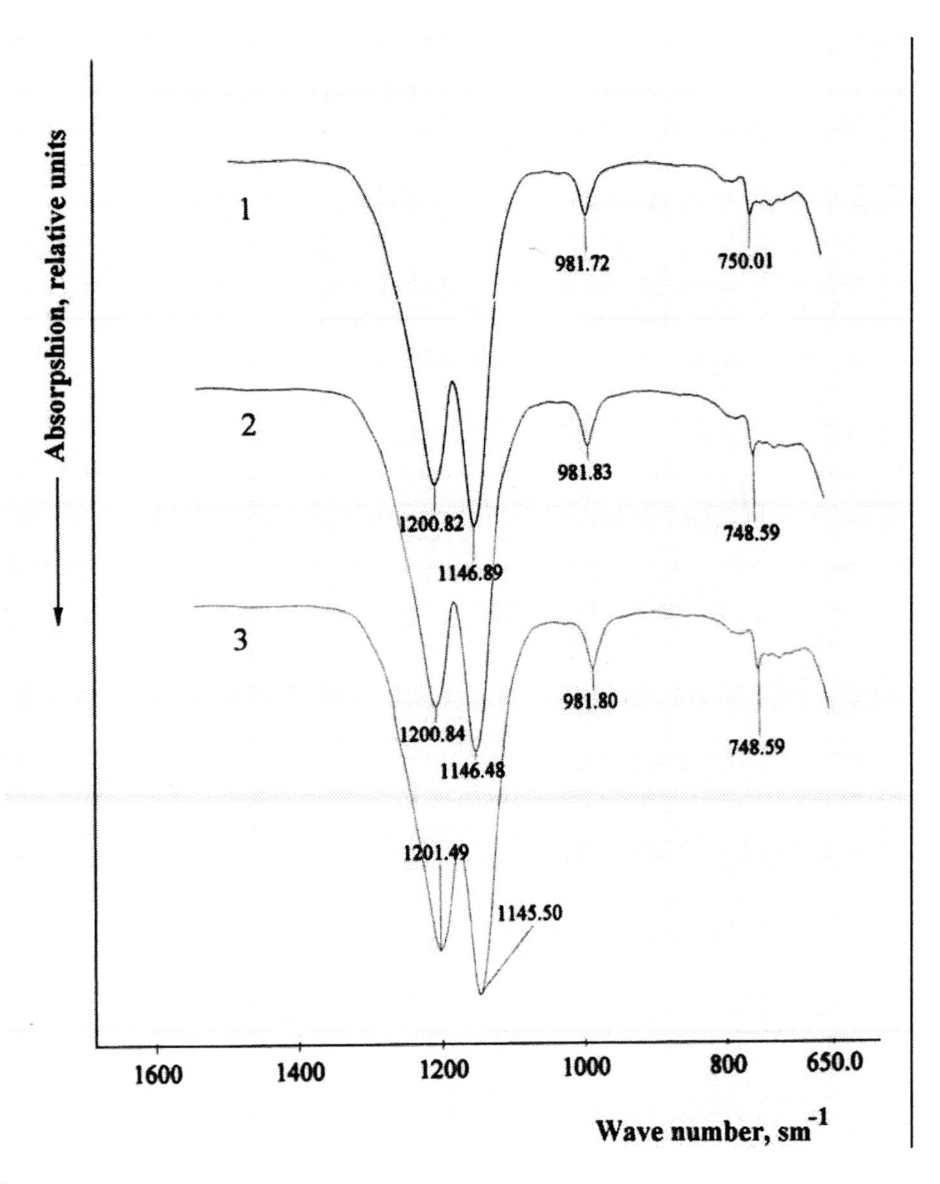

Figure 2.

The proposed solution allows to form antiadhesive thermoresistant coatings adequate to the modern requirements of the food-processing industry, and by combination of their consumer properties conforming to the best foreign samples.

REFERENCES

[1] Du Pont de Nemours E.I. , *Pat. GB 1557230*, (1979).
[2] Andreichikova, T.E., *Pat. RU 2170286*, (2001).
[3] Ladovskaya, A.A., *Pat. RU 2174137*, (2001).
[4] Yacenko, E.A., Guziy ,V.A., Ryabova, A.V., *Pat. RU 2181789*, (2002).
[5] Kudryavcev, U.D., Bespalova, Zh. I., Pyaterko, I.A., *Pat. RU 2182914*, (2002).
[6] Du Pont de Nemours E.I., *Pat. US 6518349*, (2003).
[7] Bespalova, Zh. I., Mamaev, S.A., Miroshnichenko, L.D., Kudryavcev, U.D., *Pat. RU 2202576,* (2003).
[8] Panshin ,U.A., Malkevich, S.T., Dunaevskaya, C.S., 1978: *Ftoroplasts, Chemistry,* Leningrad, p. 232.
[9] Edited by Wall ,L.A.,1972: *Fluoropolymers, Polymer Chemistry* Section National Bureau of Standards, A Division of John Wiley and Sons, Inc., New York-London-Sydney-Toronto, p.439.

In: Reactions and Properties of Monomers and Polymers ISBN 1-60021-415-0
Editors: A. D'Amore and G. Zaikov, pp. 205-215

Chapter 11

P-PARAMETER AS AN OBJECTIVE CHARACTERISTICS OF ELECTRONEGATIVITY

G. A. Korablev* and G. E. Zaikov

Basic research-educational center of chemical physics and mesoscopy
UdRC UrD RAS, Russia, Izhevsk, 426000, Institute of biochemical physics after N.M. Emmanuel RAS, , 4, Kosygina St., Moscow, 119991, Russia

ABSTRACT

It is shown that electronegativity of atoms and structures equals numerically the effective averaged energy of one valence electron for valence-active orbitals of atoms. The method for estimating electronegativity of elements and metal structures with the help of the notion of spatial-energy parameter (P-parameter) is proposed. The results of numerous computations are in accordance with reference and experimental data.

Keywords: Spatial-energy parameter, molecular and metal electronegativity.

INTRODUCTION

The notion of electronegativity was introduced in 1932 by Poling as a numerical characteristic of atoms' ability to attract electrons in a molecule. At present, a lot of methods exist to calculate the electronegativity and there are several scales of electronegativities. Discussions around the notion of electronegativity and methods for measuring it have not given single-meaning answers so far, though the notion of electronegativity is widely applied in chemical and crystal-chemical researches. It should be mentioned that lately, for instance, the tendency for using the notion of electronegativity to evaluate isomorphism processes has been observed [1,2,3].

---

* E-mail:korablev@udm.net

Since thermal-chemical, spectroscopic and other methods give consistent results, this allowed S.S. Batsanov [2] to develop the system of some averaged recommended values of electronegativity. Let us show that P-parameter can also be successfully applied as an objective characteristic of electronegativity.

Initial Regulations

The analysis of a number of physical and chemical regularities (including Lagrangian equations) shows that when heterogeneous or differently charged subsystems interact, their resulting energy can be found when adding inverse values of energy components.

Therefore, the following equations [4] were applied to the system orbital – atom nucleus:

$$\frac{1}{q^2/r_i} + \frac{1}{W_i n_i} = \frac{1}{P_E} \tag{1}$$

$$\frac{1}{q^2} + \frac{1}{(Wrn)_i} = \frac{1}{P_0} \tag{2}$$

where: W_i – orbital energy of electrons [5]; n_i – number of electrons on this orbital; r_i – orbital radius [6]; q^2/r_i – effective energy of atom nucleus, where:

$$q = \frac{Z^*}{n^*} \tag{3}$$

Here Z^* and n^* – effective nucleus charge and effective main quantum number – consider both screening effects and effects of mutual overlapping of orbitals in an atom [7];

P_E – effective spatial-energy parameter (has energy dimensionality) and in physical sense equals the energy of atom valence orbitals responsible for interatomic interactions;

P_0 – tabulated parameter, constant for given valence orbitals in each element atom – calculated based on equation (2) for many elements. Table 1 contains such computations only for the first valence electron. Values of P_0-parameters for other valence electrons are taken according to [4].

Table 1. Computation of molecular electronegativity.

Atom	Orbitals	W (eV)	r_i (Å)	q^2 (eVÅ)	P_0 (eVÅ)	R_K (Å)	P_0/3Rn (eV)	X(Batsanov)
1	2	3	4	5	6	7	8	9
Li	$2S^1$	5.3416	1.586	5.8902	3.475	1.33	0.87	0.98
Be	$2S^1$	8.4157	1.040	13.159	5.256	1.13 (M)	1.55	1.52
B	$2P^1$	8.3415	0.770	21.105	4.965	0.81	2.04	2.03
C	$2P^1$	11.792	0.596	35.395	5.868	0.77	2.54	2.56
N	$2P^1$	15.445	0.4875	52.912	6.5903	0.74	2.97	3.05
O	$2P^1$	17.195	0.4135	71.383	6.4660	0.66	3.27	3.42
F	$2S^22P^5$				50.809	0.64	3.78	3.88
Na	$3S^1$	4.9552	1.713	10.058	4.6034	1.54	1.00	0.98
Mg	$3S^1$	6.8859	1.279	17.501	5.8588	1.6 (M)	1.22	1.28
Al	$3P^1$	5.7130	1.312	26.443	5.8401	1.26	1.55	1.57
Si	$3P^1$	8.0848	1.068	29.377	6.6732	1.17	1.90	1.89
P	$3P^1$	10.659	0.9175	38.199	7.7862	1.10	2.36	2.19
S	$3P^1$	11.901	0.808	48.108	8.432	1.04	2.57	2.56
Cl	$3P^1$	13.780	0.7235	59.844	8.546	1.00	2.85	2.89
K	$4S^1$	4.0130	2.162	10.993	4.8490	1.96	0.83	0.85
Ca	$4S^1$	5.3212	1.690	17.406	5.9290	1.74	1.14	1.08
Sc	4S1	5.7174	1.570	19.311	6.1280	1.44	1.42	1.39
Ti	4S1	6.0082	1.477	20.879	6.227	1.32	1.57	1.62
V	4S1	6.2755	1.401	22.328	6.3077	1.22 1.34	1.72 1.57	(1.7) 1.54
Cr	4S23d1				17.168	1.19	1.60	1.63
Mn	4S1	6.4180	1.278	25.118	6.4180	1.18	1.81	1.73
Fe	4S1	7.0256	1.227	26.572	6.5089	1.17 1.26	1.85 1.72	1.82 1.74
Co	4S1	7.2770	1.181	27.983	6.5749	1.16	1.89	1.88

Table 1. (Continued).

Atom	Orbitals	W (eV)	r_i (Å)	q^2 (eVÅ)	P_0 (eVÅ)	R_K (Å)	$P_0/3Rn$ (eV)	X(Batsanov)
Ni	4S1	7.5176	1.139	29.348	6.6226	1.15	1.92	1.98
Cu	4S1 4S13d1	7.7485	1.191	30.117	7.0964 13.242	1.31 1.31	1.81 1.68	(1.8) 1.64
Zn	4S1	7.9594	1.065	32.021	6.7026	1.31	1.71	1.72
Ga	4P1 4S24P1	5.6736	1.254	34.833	5.9081 20.760	1.25 1.25	1.58 1.82	(1.7) 1.87
Ge	4P1 4S24P2	7.819	1.090	41.372	7.0669 30.370	1.22 1.22	1.93 2.07	2.08
As	4P1	10.054	0.9915	49.936	8.275	1.21	2.28	2.23
Se	4P1	10.963	0.909	61.803	8.5811	1.17	2.44	2.48
Br	4P1	12.438	0.8425	73.346	9.1690	1.11	2.75	2.78
Rb	5S1	3.7511	2.287	14.309	5.3630	2.22	0.81	0.82
Sr	5S1	4.8559	1.836	21.224	6.2790	2.00	1.05	1.01
Y	5S1	6.3376	1.693	22.540	6.4505	1.69	1.27	1.26
Zr	5S1	5.6414	1.593	23.926	6.5330	1.45	1.50	1.44
Nb	5S1	5.8947	1.589	20.191	6.3984	1.34	1.52	1.56
Mo	5S1	6.1140	1.520	21.472	6.4860	1.29	1.68	1.73
Tc	5S1	6.2942	1.391	30.076	6.7810	1.27	1.78	1.76
Ru	5S1	6.5294	1.410	24.217	6.6700	1.24	1.79	1.85
Ru (II)	5S14d1				15.670	1.24	2.11	(2.1)
Rh	5S1	6.7240	1.364	33.643	7.2068	1.25	1.92	1.96
Rh (5S14d	5S1	6.7240	1.364	25.388	6.7380	1.25	1.87	(1.82)
Pd	5S24d2				30.399	1.28	1.98	1.95
Pd	5S1	6.9026	1.325	35.377	7.2670	1.28	1.89	1.95; 1.85
Ag* (5S24d	5S1	7.0655	1.286	37.122	6.9898	1.25	1.86	1.90
Ag (5S14d	5S1	7.0655	1.286	26.283	6.7520	1.34	1.68	(1.66)
Cd	5S1	7.2070	1.184	38.649	6.9898	1.38	1.69	(1.68); 1.62
Jn	5S25P1				21.841	1.42	1.71	1.76; 1.68

Table 1. (Continued).

Atom	Orbitals	W (eV)	r_i (Å)	q^2 (eVÅ)	P_0 (eVÅ)	R_K (Å)	P_0 /3Rn (eV)	X(Batsanov)
Sn	5P1	7.2124	1.240	47.714	7.5313	1.40	1.79	(1.80); 1.88
Sb	5S25P3				42.502	1.39	2.04	1.98
Te	5P1	9.7907	1.087	67.286	9.1894	1.37	2.24	(2.3); 2.16
J	5P1	10.971	1.0215	77.651	9.7936	1.35	2.42	2.42
Cs	6S1	3.3647	2.518	16.193	5.5628	2.29	0.81	0.78
Ba	6S1	4.2872	2.060	22.950	6.3768	2.07	1.03	0.98
La	6S1	4.3528	1.915	34.681	6.7203	1.76	1.27	(1.18); 1.20
La	6S2	4.3528	1.915	34.681	11.259	1.69	1.11	(1.18); 1.20
Hf	6S1	5.6383	1.476	33.590	6.7151	1.44	1.55	1.54
Ta	6S1	5.9192	1.413	36.285	6.7971	1.34	1.69	(1.67); 1.57
W	6S1	6.1184	1.360	38.838	6.8528	1.30	1.76	(1.79); 1.78
Re	6S25d5	6.2783	1.310		50.867	1.28	1.89	(1.88); 1.90
Os	6S1	6.4995	1.266	42.620	7.7344	1.26	2.05	1.98
Jr	6S1	6.6788	1.227	44.655	7.7691	1.27	2.04	2.02
Pt	6S25d2				31.949	1.30	2.04	1.98; 2.05
Au	6S1	6.9820	1.187	47.849	7.0641	1.34	1.76	1.68
Au	6S1	6.9820	1.187	47.849	7.0641	1.24	1.90	1.97
Hg	½ (6S2+ 6S15d1)				15.119	1.44	1.75	1.80; 1.84
Tl (I)	6P1	5.2354	1.319	60.054	6.1933	1.44	1.43	1.40
Tl	6S26P1				22.012	1.44	1.70	1.72
Pb	6S26P2				32.526	1.50	1.81	1.87
Bi	6P1	8.7076	1.2125	71.171	9.0406	1.50	2.01	(2.03); 1.95
Po	6P1	9.2887	1.1385	80.881	9.3523	1.50	2.08	(2.2); 2.12
At	6P1	10.337	1.0775	91.958	9.9074	(1.39)	2.38	2.30

Computations and Comparisons

Due to the physical sense of P_0-parameter we assume that electronegativity for the lowest stable oxidation degree of an element equals the effective averaged energy of valence electron:

$$X = \frac{\sum Po}{3nR} \qquad (4)$$

or – for the first valence electron:

$$X = \frac{Po}{3R} \qquad (4a)$$

Here: $\sum P_0$ – total of P_0-parameters for n-valence electrons, R – atom radius (depending upon the bond type – metallic, crystalline or covalent), and the value of P_0–parameter is calculated via the bond energy of electrons as in [5]. Digit 3 in denominator of equation (4) reflects the fact that probable interatomic interaction is considered only along the bond line, i.e. in one out of three spatial directions. The calculation of molecular electronegativity for all elements by equations (4) and (4a) is given in Table 1, for metallic – in Table 2.

For some elements (characterized by the availability of both metallic and covalence bonds) the calculations of electronegativity is done in two variants – using the values of atomic and covalence radii.

Equation (4) cannot be applied to the elements of zero group, since in this case the notions of electronegativity and covalence radius for inert gases practically lose their meanings. Deviations in the calculations by equation (4) from those generally accepted values of electronegativity according to Batsanov [2,8,9] and Allred-Rokhov [9] do not exceed (2-5)% in most cases.

Thus, simple correlation (4) rather satisfactorily estimates the electronegativity value within its values according to Batsanov's and Allred-Rokhov's data.

The advantages of such an approach are wide capabilities of P-parameter to determine electronegativity of groups and compounds as P-parameter can be relatively easily calculated (based on initial rules) both for simple and complex compounds.

At the same time, it is possible to take into account individual features of structures and, consequently, not only to characterize but also to predict important physical and chemical properties of these compounds (isomorphism, mutual solubility, temperature of eutectics, etc). For instance, with the help of the notion of P-parameter it is possible to determine the electronegativity both of metallic (X_M) and crystalline (X_{CR}) structures.

In [2,8] values of electronegativity are obtained by averaging thermochemical and geometric values that, in turn, are calculated by thermochemical data for these structures and with the help of covalence and atomic radii taking bond repetition factor into consideration.

The values of X_{CR} calculated by equation (4) are also in satisfactory accordance with the corresponding values by [8] – the Table is not given.

Table 2. Computation of metal electronegativity (X_M).

Element	Orbitals	P_0(eVÅ)	n	R, R_{CR}Å	X_M (eV) (computation)	X_M (Batsanov)	X_M (O-P)
1	2	3	4	5	6	7	8
Li	$2S^1$	3.475	1	1.46 (cr) 1.34 (к)	0.79 0.86	0.92	0.88
Be	$2S^2$	7.5120	2	0.98	1.28	1.34	1.20
B	$2P^1$	4.9945	1	0.95	1.75	1.76	1.58
C	$2P^2$	10.061	2	0.77	2.18	2.18	2.00
N	$2P^3$	15.830	3	0.70	2.51	2.67	2.55
O	$2P^1$	6.4663	1	0.66	3.27	3.16	3.10
Na	$3S^1$	4.6034	1	1.66	0.92	0.98	0.93
Mg	$3S^1$	5.8588	1	1.60	1.22	1.22	1.09
Al	$3P^1$	5.840	1	1.43	1.36	1.47	1.30
Si	$3P^2$	10.876	2	1.17	1.55	1.68	1.43
P	$3P^3$ $3S^23P^3$	16.594 35.644	3 5	1.10 1.30	1.68 1.83	1.92	1.66
S	$3P^2$	13.740	2	1.04	2.20	2.17	1.88
K	$4S^1$	4.8490	1	1.96 (к) 2.08 (cr)	0.82 0.77	0.83	0.86
Ca	$4S^1$	5.929	1	1.83	1.08	0.99	0.98
Sc	$4S^1$	6.1279	1	1.64	1.24	1.14	1.12
Ti	$4S^13d^1$	11.785	2	1.46	1.35	1.27	1.26
V	$4S^1$	6.3077	1	1.34	1.40	1.41	1.42
Cr	$4S^23d^1$	17.168	3 4	1.27	1.5	1.44	1.48
Mn	$4S^13d^1$	12.924	1	1.30	1.66	1.60	1.63
Fe	$4S^1$	6.5089	1	1.26	1.72	1.68	1.72
Co	$4S^1$	6.5749	1	1.25	1.75	1.77	1.80
Co	$4S^13d^1$	12.707	2	1.17	1.81	1.77	1.80
Ni	$4S^13d^1$	12.705	2	1.16	1.83	1.85	1.89

Table 2. (Continued).

Element	Orbitals	P_0(eVÅ)	n	R, R_{CR}Å	X_M (eV) (computation)	X_M (Batsanov)	X_M (O-P)
Cu	$3d^1$	6.1457	1	1.35	1.52	1.51	1.54
Cu	$4S^2$	11.444	2	1.28	1.49	1.51	1.54
Zn	$4S^1 3d^1$	12.818	2	1.39	1.54	1.58	1.55
Ga	4S24P1	20.760	3	1.39	1.64	1.67	1.59
Ge	4P1	7.0669	1	1.39	1.69	1.79	1.66
As	4P3	18.645	3	1.18	1.76	1.88	1.70
Se	4P2	15.070	2	1.18	2.15	2.14	2.01
Rb	5S1	5.3630	1	2.22	0.81	0.81	0.85
Sr	5S1	6.2790	1	2.15	0.97	0.95	0.94
Y	5S1	6.4505	1	1.81	1.19	1.17	1.11
Zr	5S1	6.5330	1	1.60	1.36	1.30	
Zr	5S24d2	23.492	4	1.60	1.22		1.25
Nb	5S14d4	30.607	5	1.45	1.41	1.37	1.34
Mo(VI)	5S14d5	38.808	6	1.40	1.54	1.48	1.46
Tc	5S1	6.7810	1	1.36	1.66	1.56	1.55
Ru	5S1	6.6700	1	1.34	1.64	1.65	1.63
Rh	5S1	6.7378	1	1.34	1.68		1.69
Rh*	5S1	7.2068	1	1.34	1.77	1.72	
Pd	5S1	7.2670	1	1.37	1.77	1.76	1.71
Ag	5S1	6.7520	1	1.44 1.53	1.56 1.47	1.51	1.46
Cd	5S1	6.9898	1	1.56 1.48	1.49 1.57	1.57	1.46
Jn	5S25P1	21.841	3	1.66 1.44	1.46 1.68	1.64	1.47
Sn	5P1	7.5313	1	1.58 1.40	1.59 1.79	1.74	1.54

Table 2. (Continued).

Element	Orbitals	P_0(eVÅ)	n	R, R_{CR}Å	X_M (eV) computation)	X_M (Batsanov)	X_M (O-P)
Sb	5P1	8.9676	1	1.61	1.86	1.83	
	5P3	20.509	3	1.36	1.67		1.57
Te	5P1	9.1894	1	1.70	1.80		1.78
	5P2	16.170	2	1.32	2.04	2.04	
Cs	6S1	5.5628	1	2.29 (к)	0.81	0.80	0.84
Ba	6S1	6.3768	1	2.21	0.96	0.94	0.93
La	6S1	6.7203	1	1.87	1.20	1.14	
	6S2	11.259	2	1.87	1.00		1.03
Hf	6S1	6.7151	1	1.59	1.41	1.45	
	6S25d2	24.498	4	1.49	1.37		1.35
Ta	6S1	6.7971	1	1.46	1.55	1.59	1.52
W	6S1	6.8528	1	1.40	1.63		1.65
W	6S25d3	34.828	5	1.33	1.75	1.71	
Re	6S1	6.8483	1	1.30	1.76	1.80	1.74
Re	6S25d5	50.867	7	1.37	1.77	1.80	1.74
Os	6S1	7.7344	1	1.35	1.91	1.88	
Os	6S25d6	7.7344	8	1.35	1.81		1.83
Jr	6S1	7.7691	1	1.35	1.92	1.94	
Jr	6S25d2	30.790	4	1.35	1.90		1.87
Pt	6S1	7.0718	1	1.31	1.80		1.83
Pt	6S25d2	31.949	4	1.38	1.93	1.91	
Au	6S1	7.0641	1	1.44	1.64	1.66	
Au	6S1	7.0641	1	1.52	1.55		1.58
Hg	6S1	6.8849	1	1.49	1.54		1.53
Tl	6S26P1	22.012	3	1.46	1.68	1.75	1.53
Pb	6P2	13.460	2	1.46	1.54		1.53
	6S26P2	32.526	4	1.46	1.86	1.80	
Bi	6P3	21.919	3	1.41	1.73	1.86	
	6S26P3	43.969	5	1.82	1.61		1.54
Po	6P2	16.767	2	1.40	2.00	2.03	
	6S26P4	53.012	6	1.76	1.67		1.68

The capabilities of P-parameter methodology for determining the electronegativity (X) of atoms of radical groups in molecules are indicative. In these cases the value of dimensional characteristic R in equation (4) can apparently be determined by the difference of bond length (d) and covalence radius (R_K) of the second element:

$$R = d - R_K$$

Table 3 contains some calculations using the data only according to [2,10] for carbon atoms in radicals of CH_3 and CG_3 (where G – halogen) confirming the initial regulations. Out of all the data obtained it can be concluded that the effective averaged P_E-parameter of one valence electron is a direct characteristic of electronegativity. But this also means that there cannot be a single-meaning value of electronegativity of the given structure. Its value depends upon the structural interactions of values of interatomic distances changing in dynamics and upon the type of registered valence-active orbitals of bond atoms. Characteristic examples – in Pd, Rh (Table 1) and in P, Sn, Sb (Table 2) and others.

Due to the established physical sense of electronegativity it can be assumed that in prognostic purposes in many cases it is more rational not to use the averaged energy characteristic for a single valence electron, but effective energy of all valence-active electrons, i.e. the value of P_E-parameter of each valence atom orbital. In particular, this refers to those interactions having mainly covalence or ionic-covalence character of chemical bond. This is a wide range of inter-structural interactions of exchange-substitute type (solid solutions) and the processes of formation of new chemical compounds during polymerization, complex-formation, etc.

In this regard P-parameter methodology is successfully used to evaluate the degree of structural interactions during isomorphism (extension of solid solutions) to determine the directedness of processes in complex polymeric structures, etc [4].

Table 3. Computation of electronegativity of carbon atoms in radical groups.

Radical	Carbon		d (Å) by [10]	R_K (Å) of 2nd element	X (eV) (computation)	X by [2.10]
	Orbitals	P_0 (eVÅ)				
CH_3	$2P^1$	5.8680	1.12	0.28	2.3	2.4
CF_3	$2P^1$	5.8680	1.32	0.71	3.2	3.1
CCl_3	$2S^22P^2$	24.585	1.76	1.00	2.7	2.7
CBr_3	$2S^22P^2$	24.585	1.94	1.14	2.6	2.6

CONCLUSIONS

(1) P_0-parameter of free atoms is a tabulated constant of spatial-energy characteristic of structural interactions.

(2) Effective P_E-parameter averaged for a single valence electron equals the value of electronegativity of the given structural formation.

REFERENCES

[1] V.M. Yakovlev. A new method for estimating electronegativity of elements. *Journal of inorganic chemistry,* 2002.-V.47-No. 10.-P.1644-1646.

[2] S.S. Batsanov. Structural chemistry. Facts and dependencies. M.:*MSU*.-2000, 292p.

[3] M.A. Shumilov. On conditions of unlimited mutual solubility of metals in solid state. *News of HEIs – Ferrous metallurgy*. -2001.-No. 10.-P.19-21.

[4] Korablev G.A. *Spatial-Energy Principles of Complex Structures Formation*, Netherlands, Brill Academic Publishers and VSP, 2005,426p. (Monograph).

[5] Fischer C.F. Average-Energy of Configuration Hartree-Fock Results for the Atoms Helium to Radon. *Atomic Data*,-1972, -No. 4, -p. 301-399.

[6] Waber J.T., Cromer D.T. Orbital Radii of Atoms and Ions. *J. Chem. Phys* -1965, -V 42, -No. 12, -p. 4116-4123.

[7] Clementi E., Raimondi D.L. Atomic Screening constants from S.C.F. Functions, 1. *J.Chem. Phys*.-1963, -v.38, -No. 11, -p. 2686-2689.

[8] S.S. Batsanov. System of electronegativities and effective charges of atoms for crystalline compounds. *Journal of inorganic chemistry*, 1975, No. 10.-P.2595-2600.

[9] S.S. Batsanov, R.A. Zvyagina. Lapping integrals and problem of effective charges. *Nauka: Novosibirsk*, 1966.- 386p.

[10] S.S. Batsanov. Structural refractometry. *M.: Vysshaya shkola*, 1976, 304p.

INDEX

A

B

C

D

E

F

G

H

I

N

O

P

Q

T

U

V

W

X

Y

Z